云计算技术与应用专业系列教材

OpenStack 云计算基础架构平台技术与应用

沈建国 陈永 主编
余立强 杜纪魁 刘卓华 副主编
杨国华 史宝会 主审

人民邮电出版社
北京

图书在版编目（CIP）数据

OpenStack云计算基础架构平台技术与应用 / 沈建国，陈永主编. -- 北京 : 人民邮电出版社，2017.1（2022.6重印）
云计算技术与应用专业系列教材
ISBN 978-7-115-44541-4

Ⅰ. ①O… Ⅱ. ①沈… ②陈… Ⅲ. ①云计算一研究 Ⅳ. ①TP393.027

中国版本图书馆CIP数据核字(2016)第326794号

内 容 提 要

本书较为全面地介绍了开源的 OpenStack 云计算架构及其组件，并借助开源脚本搭建形成一个完整的云平台。全书共分为认识 OpenStack、环境设计和系统准备、认证服务、基础控制服务、网络服务、虚拟化服务、存储服务、高级控制服务和平台构建脚本解读等 9 个项目。

本书可以作为云计算技术与应用专业、计算机网络技术专业及其他计算机相关专业的云计算课程教材，也可以作为云计算相关的培训班教材，还可供云计算相关从业人员和广大计算机爱好者自学使用。

◆ 主　　编　沈建国　陈　永
副 主 编　余立强　杜纪魁　刘卓华
主　　审　杨国华　史宝会
责任编辑　桑　珊
执行编辑　左仲海
责任印制　焦志炜

◆ 人民邮电出版社出版发行　　北京市丰台区成寿寺路 11 号
邮编　100164　　电子邮件　315@ptpress.com.cn
网址　http://www.ptpress.com.cn
三河市君旺印务有限公司印刷

◆ 开本：787×1092　1/16
印张：16.5　　　2017 年 1 月第 1 版
字数：399 千字　　　2022 年 6月河北第12次印刷

定价：42.00 元

读者服务热线：(010)81055256　印装质量热线：(010)81055316
反盗版热线：(010)81055315

序 PREFACE

云计算被称为继个人计算机、互联网之后的第三次信息技术革命，基于云计算的应用服务已经成为全球高科技产业竞争的前沿领域。

2015年1月，国务院印发《关于促进云计算创新发展培育信息产业新业态的意见》，在关于加强专业人才培养中提出："鼓励普通高校、职业院校、科研院所与企业联合培养云计算相关人才，加强学校教育与产业发展的有效衔接，为云计算发展提供高水平智力支持。"

教育部已于2015年10月公布了高职高专专业目录，新增了"云计算技术与应用"专业，截至2016年，全国约50所高职院校开设了"云计算技术与应用"专业。

2015年11月，工业和信息化职业教育教学指导委员会领导，组织江苏省教育厅、山东省教育厅、高等教育出版社、南京第五十五所技术开发有限公司，以及国内部分高职院校的领导和专家成立了"云计算技术与应用专业教材编审委员会"。编审委员会专家成员单位由北京信息职业技术学院、深圳职业技术学院、上海电子信息职业技术学院、常州信息职业技术学院、南京信息职业技术学院、重庆电子工程职业学院、无锡商业职业技术学院、无锡科技职业学院、山东商业职业技术学院、贵州交通职业技术学院、湖南工程学院、北京青年政治学院、湖南铁道职业技术学院、南京第五十五所技术开发有限公司、高等教育出版社、人民邮电出版社等组成。

"云计算技术与应用专业教材编审委员会"组织相关专家编写《OpenStack云计算基础架构平台技术与应用》《Hadoop大数据平台构建与应用》《虚拟化技术与应用》《Android云应用移动客户端开发》《云计算存储技术与应用》《云计算网络技术与应用》和《云计算数据中心运维》等云计算技术与应用专业系列教材。本书是其中的一本，是在"编审委员会"的组织、设计和指导下，由成员单位无锡商业职业技术学院和南京第五十五所技术开发有限公司共同编写完成的。

希望本书的所有读者，在了解到云计算技术的同时，都能够积极地投身到云计算产业实践中来。只有更多的人认识到云计算的价值，云计算产业才会有源源不断的发展动力，相信读者中会有很多人成为云计算产业的中坚力量。

云计算技术与应用专业教材编审委员会

2016年11月

前言 FOREWORD

计算机技术经历了从大型主机、个人计算机、客户/服务器计算模式到今天的互联网计算模式的演变，尤其是互联网 Web 2.0 技术的应用，使计算能力需求更多地依赖于通过互联网连接的远程服务器资源。作为资源的提供者，需要具备超高的计算性能、海量的数据存储能力、网络通信能力和随时的扩展能力。在多种应用需求的推动下催生了虚拟化技术和云计算技术。当今，云计算技术已经发展形成信息技术应用服务平台、云存储技术、大数据分析、互联网+技术等基础平台，在信息技术的发展过程中起着平台支撑作用。

云计算是推动信息技术能力实现按需供给，促进信息技术和数据资源充分利用的全新业态，是信息化发展的重大变革和必然趋势。发展云计算，有利于分享信息知识和创新资源，降低全社会创业成本，培育形成新产业和新消费热点，对稳增长、调结构、惠民生和建设创新型国家具有重要意义。

为适应高职院校对云计算技术专业教学的需求，本书在“云计算技术与应用专业教材编审委员会”的组织和指导下，由成员单位无锡商业职业技术学院和南京第五十五所技术开发有限公司共同编写。本书是校企产教融合后的实践产物，基于开源的 OpenStack 云计算架构，解决高职高专院校云计算专业或相关专业的云计算架构搭建与应用的教学需求。遵循以项目为驱动、任务为目标的编写思路，每个项目分为若干个子任务，每个任务分成 3 部分进行。第 1 部分提出具体的任务要求，第 2 部分讲解任务的相关知识，第 3 部分介绍完成任务的具体操作步骤，做到基础知识介绍具有针对性，任务目标操作具体化。书后还提供了所有搭建脚本的附录，以方便读者查阅。

本书的参考学时为 52 ~ 70 学时，建议采用理论实践一体化教学模式，各项目的参考学时见下面的学时分配表。

学时分配表

项 目	课程内容	学 时
项目一	认识 OpenStack	6
项目二	环境设计和系统准备	8
项目三	认证服务	6
项目四	基础控制服务	8
项目五	网络服务	8
项目六	虚拟化服务	6
项目七	存储服务	6
项目八	高级控制服务	8
项目九	平台构建脚本解读	6
	课程考评	2
课时总计		64

本书由沈建国、陈永任主编，余立强、杜纪魁、刘卓华任副主编，杨国华、史宝会任主审，张云参与了部分章节的编写工作。南京第五十五所技术开发有限公司工程师参与了本书的案例设计和案例测试，在此表示衷心的感谢。

本书配套的配套资源、演示文档等，可登录人邮教育社区（http://www.ryjiaoyu.com）下载。

虽然本书编者已尽可能做到更好，但由于搭建环境的复杂性，书中疏漏和不足之处在所难免，殷切希望广大读者批评指正。同时，恳请读者一旦发现错误，于百忙之中及时与编者联系，以便尽快更正，编者将不胜感激，E-mail：shenjianguo@wxic.edu.cn。

编 者

2016 年 11 月

目 录 CONTENTS

项目一 认识 OpenStack

在本书开始编写之时，OpenStack 已经发布了自己的第 13 个版本——Mitaka，而距 OpenStack 发布第一个版本——Austin 项目也才仅仅 6 年，可见 OpenStack 的发展是非常迅速的，当然它的发展离不开各大厂商的支持，也受当前社会经济发展驱动。下面，我们就开始揭开神秘的 OpenStack 的面纱。

学习目标

- 了解云计算的基本概念。
- 理解云计算与虚拟化的基本概念。
- 认识 OpenStack。

任务一 初识云计算

任务要求

小李刚从学校毕业，被某公司聘为云计算助理工程师。公司现准备将原有的计算机服务器改造成云计算服务平台。为此，小李必须去了解云计算的基础概念及搭建云计算平台的相关知识，以便提出详细的改建方案和实施步骤。需要认识以下的基本知识。

- 云计算的起源。
- 云计算的基本概念。
- 典型的云计算服务平台。

相关知识

1. 云计算的起源

早在 2006 年 3 月，亚马逊公司首先提出弹性计算云服务。2006 年 8 月 9 日，谷歌公司首席执行官埃里克·施密特（Eric Schmidt）在搜索引擎大会（SES San Jose 2006）上首次提出“云计算”（Cloud Computing）的概念。从那时候起，云计算开始受到关注，这算是云计算最正统的诞生记。

云计算作为一种计算技术和服务理念，有着极其深厚的技术背景，谷歌作为搜索公司，首创这一概念，有着很大的必然性。随着众多的互联网厂商的兴起，各家互联网公司对云计算的投入和研发不断加深，陆续形成了完整的云计算技术架构、硬件网络，服务器方面逐步向数据中心、全球网络连接、软件系统等方面发展，完善了操作系统、文件系统、并行计算架构、并行计算数据库和开发工具等云计算系统关键部件。

云计算经历了集中式时代向网络时代的转变，最终向分布式时代转换，在分布式时代基础之上形成了云时代，如图 1-1 所示。

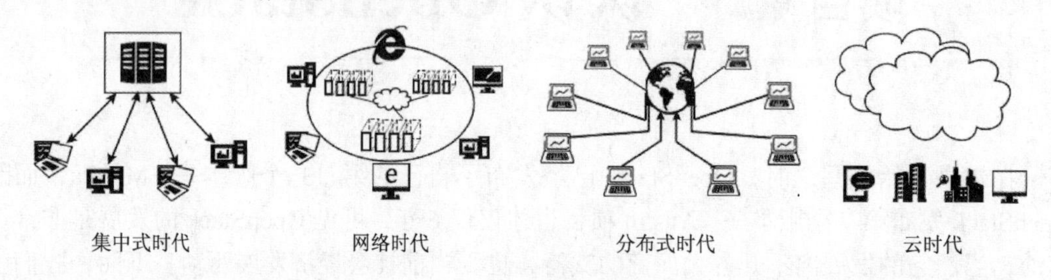

图 1-1 云计算起源

云计算的最终目标是将计算、服务和应用作为一种公共设施提供给公众，使人们能够像使用水、电、煤气和电话那样可以便捷地使用计算资源。2010 年 7 月，美国国家航空航天局（NASA）和 Rackspace、AMD、Intel、戴尔（DELL）等支持厂商共同宣布"OpenStack 开放源代码计划"。微软公司在 2010 年 10 月表示支持 OpenStack 与 Windows Server 2008 R2 的集成；Ubuntu 也已把 OpenStack 加至 11. 04 版本中。2011 年 2 月，思科系统正式加入 OpenStack，重点研制 OpenStack 的网络服务。云计算服务得到这些厂商支持，发展的速度变得更快，OpenStack 项目也得到了空前的发展，迎来了历史最好的发展时机。

任何新科技的广泛应用在帮助我们增加处理能力的同时，也同样地对云计算环境的安全性造成威胁。通过网络的计算能力，可以将原本安装在自己电脑上的软件安装到运营商提供的数据存储中心，取代我们将自己的资料存放在本地硬盘的动作，这也导致了当本地或者运营商出现暂时故障，无法使用某项服务时，短期内无法正常访问自己的服务器的问题，如果遇到严重问题时可能会遗失全部资料。如今，技术越来越成熟，对数据保护的手段越来越多，出现较大的服务访问故障现象的概率偏低。

2. 云计算的基本概念

（1）云计算的定义

根据美国国家标准与技术研究院（NIST）定义，云计算是一种按使用量付费的模式，这种模式提供可用的、便捷的、按需的网络访问，以进入可配置的计算资源共享池（资源包括网络、服务器、存储和应用软件等），这些资源能够被快速提供，只需投入很少的管理工作或与服务供应商进行很少的交互。

云计算是分布式计算技术中的一种，最基本的概念是通过网络将庞大的计算处理程序自动拆分成无数个较小的子程序，再交由服务器集群组成的庞大系统进行搜索、计算、分析之后，将处理的结果返回给用户。通过这样的计算处理，可以使用户在数秒之间处理数以万计的数据量。

（2）云计算的五大特征

① 按需自助服务：消费者可以单方面部署资源。例如，服务器、网络存储的资源是按需自动部署的，不需要与服务供应商进行人工交互。

② 通过互联网获取：资源可以通过互联网获取，并可以通过标准方式访问。例如，通

过瘦客户端或富客户端（移动电话、便携式计算机和工作站等）。

③ 资源池化：供应商的资源被池化，以便以多用户租用模式被不同客户使用。例如，不同的物理和虚拟资源可根据客户需求动态分配和重新分配，通常与地域无关，这些资源包括存储器、处理器、内存和网络带宽。

④ 快速伸缩：资源可以弹性地部署和释放，有时是自动化的，以便能够迅速地按需扩大和缩小规模。

⑤ 可计量：云计算系统通过使用一些与服务种类（存储、计算、带宽、激活的用户账号）对应的抽象信息提供计量能力（通常在此基础上实现按使用量付费）。

（3）云计算的服务模型——SPI 模型

云计算的 SPI 模型由 3 大服务组成，即 IaaS（基础设施即服务）、PaaS（平台即服务）和 SaaS（软件即服务），三者构成如图 1-2 所示。

① IaaS：消费者使用"基础计算资源"。资源服务包括处理能力、存储空间、网络组件或中间件服务。消费者能掌控操作系统、存储空间、已部署的应用程序及网络组件（如防火墙、负载均衡器等），但并不掌控云基础架构，如 Amazon AWS、Rackspace 等。

② PaaS：消费者使用主机操作应用程序。消费者掌控运作应用程序的环境（也拥有主机的部分掌控权），但并不掌控操作系统、硬件或网络基础架构。平台通常是应用程序基础架构，如 Google App Engine。

③ SaaS：消费者使用应用程序，但并不掌控操作系统、硬件或网络基础架构。它是一种服务观念的基础，软件服务供应商以租赁（而非购买）的概念提供客户服务，比较常见的模式是提供一组账号密码。

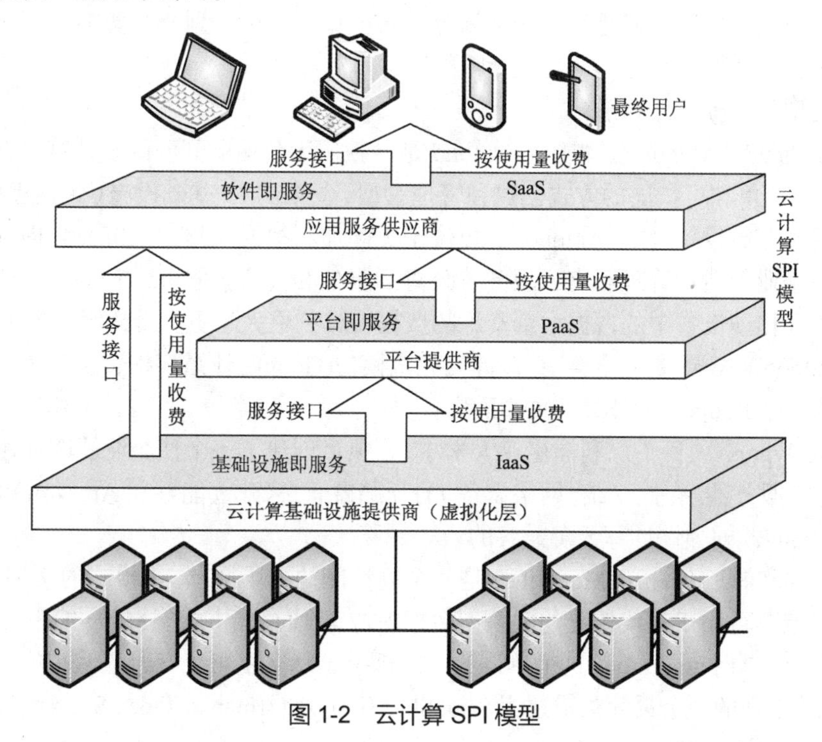

图 1-2　云计算 SPI 模型

在图 1-2 所示的模型中，IaaS 主要是对应基础设施，可实现底层资源虚拟化，以及实际云应用平台部署，这是一个网络架构由规划架构到最终物理实现的过程；PaaS 基于 IaaS 技术和平台，部署终端用户使用的应用或者程序，提供对外服务的接口或者服务产品，最终实现整个平台的管理和平台的可伸缩化；SaaS 基于现成的 PaaS 的基础，作为终端用户的最后接触产品，完成现有资源的对外服务，以及服务的租赁化。

（4）云计算的四大部署类型

① 私有云：云计算基础设施由一个单一的组织部署和独占使用，可由该组织、第三方或两者的组合来拥有和管理。

② 社区云：云计算基础设施由一些具有共有关注点的组织形成的社区中的用户部署和使用，可由一个或多个社区中的组织、第三方或两者的组合来拥有、管理和运营。

③ 公有云：云计算基础设施被部署给广泛的公众开放地使用。它可能被一个商业组织、研究机构、政府机构或者几者的混合所拥有、管理和运营，被一个销售云计算服务的组织所拥有，该组织将云计算服务销售于一般人或广泛的工业群体。

④ 混合云：云计算基础设施是由两种或两种以上的云（私有、社区或公有）组成的，每种云仍然保持独立，但用标准的或专有的技术将它们组合起来，具有数据和应用程序的可移植性。

3. 云计算平台分类

（1）云计算平台分类

从云计算平台的技术应用看，云计算平台可以划分为 3 类：以数据存储为主的存储型云平台，以数据处理为主的计算型云平台，以及计算和数据存储处理兼顾的综合云计算平台。

按构建云计算平台过程是否收费来划分，云计算平台可以划分为两类：开源云计算平台和商业化云计算平台。

（2）典型的开源云计算平台

① AbiCloud（Abiquo 公司）：AbiCloud 是一款用于公司的开源的云计算平台，使公司能够以快速、简单和可扩展的方式创建和管理大型、复杂的 IT 基础设施（包括虚拟服务器、网络、应用和存储设备等）。Abiquo 公司位于美国加利福尼亚州红木市，它提供的云计算服务包括为企业创造和管理私有云服务、公有云服务和混合云服务，它能让企业用户把他们的计算机和移动设备中的占据大量资源的数据转移到更大、更安全的服务器上。

② Hadoop（Apache 基金会）：该计划完全模仿 Google 体系架构，是一个开源项目，主要包括 Map、Reduce 和 HDFS 文件系统。

③ Eucalyptus 项目（加利福尼亚大学）：该项目创建了一个使企业能够使用它们内部 IT 资源（包括服务器、存储系统、网络设备）的开源界面，来建立能够和 Amazon EC2（Elastic Compute Cloud，即弹性计算云）兼容的云。

④ MongoDB（10gen）：MongoDB 是一个高性能的、开源的、无模式的文档型数据库，它在许多场景下可用于替代传统的关系型数据库或键/值存储方式。MongoDB 由 C++语言编写，其名字来自 humongous 这个单词的中间部分，从名字可见其野心所在，就是海量数据的处理。关于它的一个最简洁描述为 Scalable、High-performance、Open Source、Schema-free 和 Document-oriented Database。

⑤ OpenStack 项目：OpenStack 是一个由 NASA（美国国家航空航天局）和 Rackspace 合作研发的，以 Apache 许可证授权的自由软件和开放源代码项目。

OpenStack 是一个开源的云计算管理平台项目，由几个主要的组件组合起来完成具体工作。OpenStack 支持几乎所有类型的云环境，项目目标是提供实施简单、可大规模扩展、丰富、标准统一的云计算管理平台。OpenStack 通过各种互补的服务提供了基础设施，即服务（IaaS）的解决方案，每个服务提供 API 以进行集成。

（3）典型的商业化云计算平台

国内典型的商业化云计算平台有阿里云、盛大云和新浪云等，这个作为基础架构层的 IaaS，也就是他们所提供的云主机服务。另外，还有平台层的，包括腾讯的开放平台和新浪的开放平台（PaaS）。他们的概念和 Google 公司的 App Engine 相似，能让更多的开发者上去做应用，都是基于 Apple 的 App Store 的成功商业模式。相比于国外应用层的服务，国内应用层（SaaS）还需要走很大一段路。

国外典型的商业化云计算平台有微软、Google、IBM、Oracle 和 Amazon 等。这些国外云计算平台主要提供云企业服务，如微软的 Azure 平台，Google 公司的 Google App Engine 应用代管服务，IBM 公司的虚拟资源池提供的企业云计算整合方案，Oracle 的 EC2 上的 Oracle 数据库，OracleVM 和 Sun xVM，Amazon 公司的 EC2、S3（Simple Storage Service，即简单存储服务），SimpleDB 和 SQS。

提示

随着应用需求的不断提高和计算机先进技术的不断涌现，催生着新兴技术的应用和新概念的出现，可通过参观校园信息化中心，从实际应用角度去理解云计算技术的相关概念。

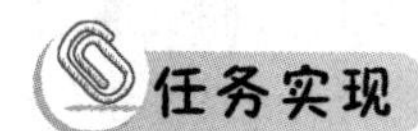

1. 参观校园信息化中心机房

通过参观校园信息化中心机房，可以比较全面地了解自己校园的整体网络拓扑结构，从而比较直观地理解上述云计算的相关概念。因此，在参观时要观察以下几点并做好记录。

① 认真听专业技术人员的情况介绍，了解本校的校园信息化建设的总体目标。

② 记录好校园信息化建设的网络拓扑结构。

③ 记录好网络拓扑结构中主要的网络设备、服务器。

④ 分辨网络服务平台是云计算架构还是虚拟化技术架构。

⑤ 在现有的系统架构中运行了哪些应用项目。

做好上述的参观和记录后再完成以下问题。

2. 分析信息化校园的网络

① 学校的网络拓扑结构是什么样的？由哪几个层次组成？

② 系统采用的是云计算平台还是虚拟化技术？并指出是什么具体的平台技术？

③ 系统有几台物理服务器？服务器的性能指标如何？

④ 系统中运行的应用项目有哪些？

⑤ 系统在运行过程中是如何保障硬件的安全及软件系统可靠运行的？对于服务器故障有什么故障修复技术？

提示

学习云计算的基本概念后，通过具体的系统环境参观，认识设备和相应的功能，并带着有关的问题去听、去观察，能够更好地理解学习内容。

任务二　虚拟化的概念

任务要求

对于入职不久的小李来说，对云计算与虚拟化的基本概念及它们之间的关系不是很清楚，借公司搭建云计算服务平台之际，要搞清虚拟化与云计算的关系。

- 虚拟化技术。
- 虚拟化技术与云计算的关系。

相关知识

1. 虚拟化技术

在图 1-2 所示的模型中，IaaS 是基础架构设施平台，可实现底层资源虚拟化，以及实际云应用平台部署。云计算、OpenStack 都离不开虚拟化的内容，因为虚拟化是云计算重要的支撑技术之一。

在计算机科学领域中，虚拟化代表着对计算资源的抽象，而不仅仅局限于虚拟机的概念。例如，对物理内存的抽象，产生了虚拟内存技术，使得应用程序认为其自身拥有连续可用的地址空间（Address Space），而实际上，应用程序的代码和数据可能被分隔成多个碎片页或段，甚至被交换到磁盘、闪存等外部存储器上，即使物理内存不足，应用程序也能顺利执行。那么，到底什么是虚拟化呢？

（1）虚拟化定义

虚拟化是一个广义的术语，是指计算元件在虚拟而不是真实的环境中运行，是一个为了简化管理、优化资源的解决方案。

在计算机运算中，虚拟化通常扮演硬件平台、操作系统（OS）、存储设备或者网络资源等角色。

图 1-3 所示为虚拟化示意图，下面从以下几个方面简单说明。

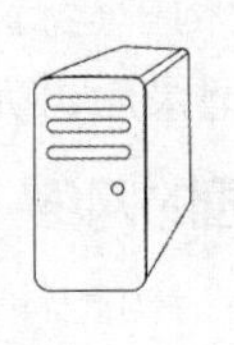

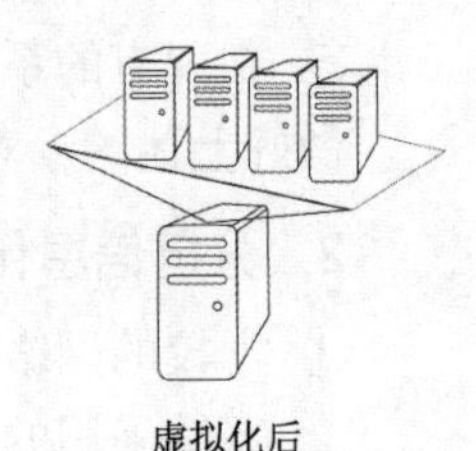

图 1-3　虚拟化示意图

① 虚拟化前：一台主机对应一个操作系统。后台多个应用程序会对特定的资源进行争抢，存在相互冲突的风险；在实际情况下业务系统与硬件进行绑定，不能灵活部署；就数据的统计来说，虚拟化前的系统资源利用率一般只有 15%左右。

② 虚拟化后：一台主机可以虚拟出多个操作系统。独立的操作系统和应用拥有独立的

CPU、内存和 I/O 资源，相互隔离；业务系统独立于硬件，可以在不同的主机之间迁移；充分利用系统资源，对机器的系统资源利用率可以达到 60%。

（2）虚拟化分类

虚拟化分类包括桌面虚拟化、应用虚拟化、服务器虚拟化等。

① 桌面虚拟化：将原本在本地电脑安装的桌面系统统一在后端数据中心进行部署和管理；用户可以通过任何设备，在任何地点、任何时间访问属于自己的桌面系统环境。

如微软的 Remote Desktop Services、Citrix 的 XenDesktop、VMware 的 View。

② 应用虚拟化：将原本安装在本地电脑操作系统上的应用程序统一运行于后台终端服务器上。用户可以通过任何设备，在任何地点、任何时间访问属于自己的应用软件。

如微软的 WTS、Citrix 的 XenApp、VMware 的 Thinapp。

③ 服务器虚拟化：将服务器物理资源如 CPU、内存、磁盘和 I/O 等抽象成逻辑资源，形成动态管理的“资源池”，并创建合适的虚拟服务器，实现服务器资源整合，提升资源利用率，最终更好地适应 IT 业务的变化。

如微软的 Hyper-V、Citrix 的 XenServer、VMware 的 ESXi。

2. 虚拟化技术与云计算的关系

云计算是很大很广泛的含义范畴，是中间件技术、分布式计算（网格计算）、并行计算、效用计算、网络存储、虚拟化和负载均衡等网络技术发展融合的产物。

虚拟化技术也不一定必须与云计算相关，如 CPU 虚拟化技术、虚拟内存等也属于虚拟化技术，但与云概念无关。云计算和虚拟化的关系如图 1-4 所示。

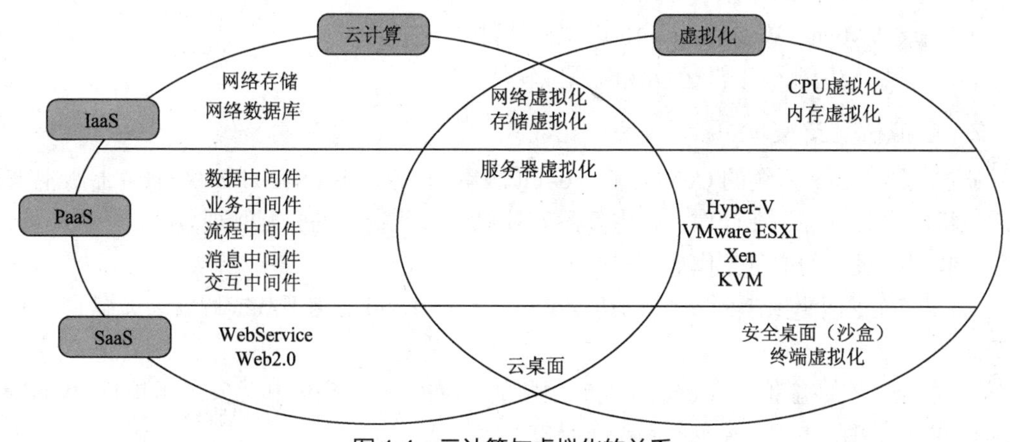

图 1-4 云计算与虚拟化的关系

3. 虚拟化技术的应用

1965 年，IBM 公司发布了最早在商业系统上实现虚拟化的产品，在此之后的 30 多年里，基于 PC 服务器的虚拟化技术发展一直很缓慢，直到 1999 年 VMware 发布 VMware Workstation 产品，之后又推出 VMware GSX Workstation（托管）和 VMware ESX Server（不托管）两款产品，该技术才顺利进入服务器市场。

VMware Workstation 可以工作在 Linux 和 Windows 操作系统上，后来 AMD 公司和 Intel 公司的处理器在内核设计中实现了硬件虚拟化的功能。紧接着，Intel 公司和 AMD 公司均在自己的产品上实现了基于硬件的虚拟化技术（Intel 公司实现的是 VT-D 技术，AMD 公司实现的是 SVM 技术）。这些技术使得虚拟化技术得到了飞速发展，主要实现了服务器虚拟化、存储虚拟化和网络虚拟化等 3 个方面的具体虚拟化技术。

提示

通过完成任务一的参观学习及查阅相关的资料，熟悉当前使用较多的 VMware 公司的虚拟化技术软件产品。有条件的可以看一下 VMware Server 的运行管理界面，也可以在个人计算机上安装 VMware Workstation 10 以上的试用版本，这对理解什么是虚拟化技术及虚拟化技术与云计算的关系有很大帮助。

任务实现

1. VMware Workstation 的安装

在用户个人计算机上安装 VMware 公司的虚拟机软件 VMware Workstation 10（试用）版本，然后再进行虚拟机安装，学习虚拟机中计算机资源的分配等相关知识，为进一步学习 VMware Server 的虚拟化管理及 VMware 的云计算服务管理打下基础。

安装虚拟机软件的准备工作如下。

① 配备好一台具有多核 CPU 的 Windows 系统计算机，内存为 4GB 或 8GB，并预留 50GB 的磁盘空间。

② 准备好 VMware Workstation 10（试用）版本的安装软件，可到 VMware 公司的官方网站去下载。

安装虚拟机软件的操作任务如下。

① 安装 VMware Workstation 10 虚拟机软件。

② 安装完成后能够正常运行该虚拟机软件。

2. 虚拟机的安装

通过安装 Linux 系统的 CentOS 6.5_64bit 版本和 Windows Server 2003 服务器版本来熟悉虚拟机的安装方法，在操作过程中熟悉计算机虚拟化资源的分配管理方法。

安装虚拟机时的准备工作如下。

① 准备好两个网络操作系统的镜像文件（*.iso）。可上网下载或用安装光盘自己制作镜像文件。

② 检查为安装虚拟机 CentOS 6.5 预留的 20GB 硬盘空间和为安装虚拟机 Windows Server 2003 预留的 15GB 硬盘空间是否足够。

启动 VMware Workstation 虚拟机软件，完成下列操作任务。

① 安装虚拟机 CentOS 6.5_64bit Linux。

② 安装 Windows Server 2003。

安装完成后分别启动两台虚拟机，并进行有关的配置、调试和应用软件的运行。

虚拟机软件为安装虚拟机提供了运行环境，从而使一台个人计算机能够虚拟出各种操作系统的虚拟机，虚拟化技术的应用给我们的教学和云计算技术的社会服务提供了很好的帮助。

提示

在 VMware Workstation 虚拟机软件中创建虚拟机时，除了要指定安装操作系统的镜像文件所在位置，最好选择“稍后安装”操作系统选项，即先把创建虚拟机的虚拟化资源都配置好，然后等想安装虚拟机操作系统时再去运行安装该虚拟机。

任务三 OpenStack 项目

任务要求

小李经过云计算及相关基本概念的学习，认为公司可以采用开源的 OpenStack 云计算解决方案搭建公司的私有云。主要考虑以下几个有利因素：首先，开源的 OpenStack 云计算解决方案可以为公司节省很多费用；其次，OpenStack 技术的发展日趋成熟，OpenStack 组件数据不断地增加，新支持的功能也在不断丰富，能够满足公司对云计算平台应用的需求；第三，小李在校期间参加过全国高职院校云计算技术与应用的职业技能竞赛，对 OpenStack 的云平台搭建比较熟悉。因此，小李需要重新认识一下 OpenStack 云计算的技术性能。

- OpenStack 的技术性能。
- IaaS 云服务商。
- OpenStack 基金会。
- OpenStack 项目调研。

相关知识

1. OpenStack 的技术性能

OpenStack 的快速发展得益于云计算技术的发展，也借助于虚拟化革命的出现。OpenStack 为一个开源的云计算解决方案，我们可以将 OpenStack 简单理解成一个开源的操作系统，它是由 Python 语言编写的，主要通过命令行（CLI）、程序接口（API）或者基于 Web 界面（GUI）实现对底层的计算资源、存储资源和网络资源的集中管理功能。在设计系统架构时可以直接运用物理硬件作为底层，我们主要将其作为基础设施即服务（IaaS）方案使用。

OpenStack 是云计算平台中的一个佼佼者，在云计算平台研发方面，国外有 IBM、微软、Google 以及 OpenStack 的鼻祖——亚马逊的 AWS 等。国内则有 Ucloud、海云捷迅、UnitedStack、EasyStack、金山云、阿里云等。现在比较流行的有 CloudStack 、Eucalyptus、vCloud Director 和 OpenStack。OpenStack 在市场中占据了绝对的份额优势。OpenStack 社区聚集着一批有实力的厂商和研发公司，他们把自己的代码贡献给社区，不断完善和推动 OpenStack 技术的发展。OpenStack 是一个云管理的项目，每年 4 月和 10 月都会有一次发布新产品的峰会。随着 OpenStack 组件的数据不断地增加，新支持的功能也在不断丰富。

（1）OpenStack 的核心项目

图 1-5 所示为 OpenStack 包含的 12 个核心项目的演变示意图。

① Austin：第一个发布的 OpenStack 项目，其中包括 Swift 对象存储和 Nova 计算模块，

有一个简单的控制台，允许用户通过 Web 管理计算和存储。

② Bexar：增加了 Glance 项目，负责镜像注册和分发；Swift 中增加了对大文件的支持和 S3 接口的中间件，在 Nova 中增加了对 RAW 磁盘格式的支持等。

③ Cactus：在 Nova 中增加了对虚拟化技术的支持，包括 LXC、Vmware、ESX；同时支持动态迁移虚拟机。

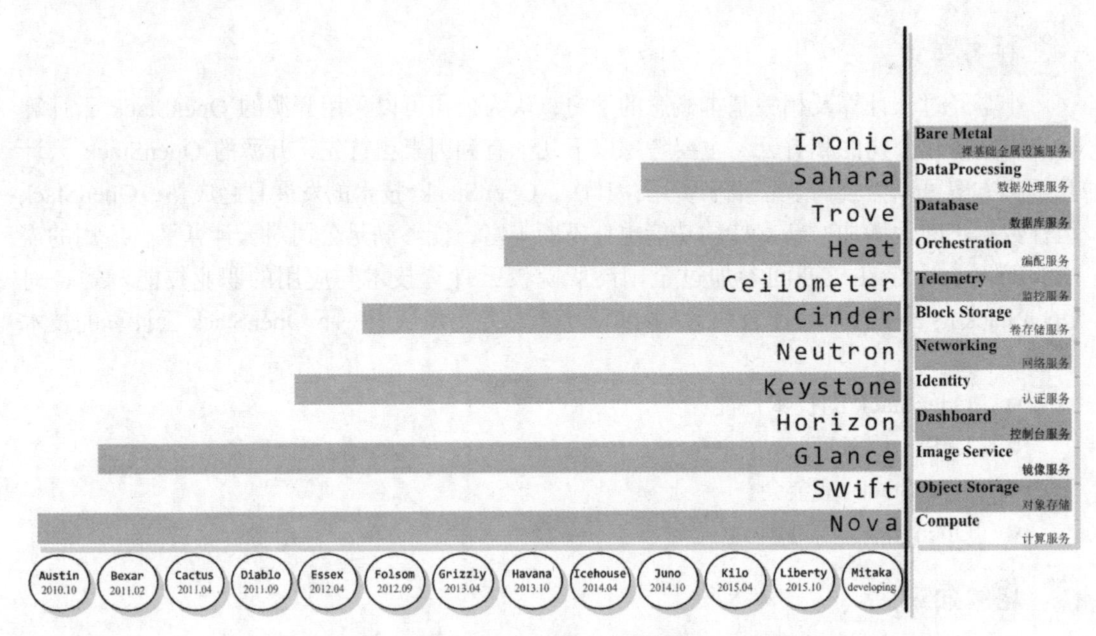

图 1-5　OpenStack 项目演变

④ Diablo：整合了 Keystone 认证，支持 KVM 的暂停和恢复、KVM 的迁移、全局的防火墙。

⑤ Essex：正式发布 Horizon，支持第三方的插件扩展 Web 控制台，发布 Keystone 项目，提供认证服务。

⑥ Folsom：正式发布 Quantum（Neutron 的前身）项目，提供网络服务；正式发布 Cinder 项目，提供块存储服务。Nova 支持 LVM 为后端的虚拟机，支持动态和块迁移等。

⑦ Grizzly：Nova 支持分布在不同的地理位置的集群组成一个 cell，支持通过 libguestfs 直接向 guest 文件系统中添加文件；通过 Glance 提供的 image 位置 URL 直接获取 image 内容以加速启动；支持无 image 条件下启动带块设备的实例；支持为虚拟机实例设置（CPU、磁盘 I/O、网络带宽）配额。

⑧ Havana：正式发布 Ceilometer 项目，进行（内部）数据统计，可用于监控报警；正式发布 Heat 项目，让应用开发者通过模板定义基础架构并自动部署；网络服务 Quantum 变更为 Neutron；Nova 中支持在使用 cell 时同一 cell 中虚拟机的动态迁移；支持 Docker 管理的容器；使用 Cinder 卷时支持加密；Neutron 中引入一种新的边界网络防火墙服务；可通过 VPN 服务插件支持 IPSec VPN；Cinder 中支持直接使用裸盘做存储设备，无需再创建 LVM。

⑨ Icehouse：新项目 Trove（DB as a Service）现在已经成为版本中的组成部分，它允许用户在 OpenStack 环境中管理关系数据库服务；对象存储（Swift）项目有比较大的更新，包括可发现性的引入和一个全新的复制过程（称为 s-sync）；联合身份验证将允许用户通过相同认证信息同时访问 OpenStack 私有云与公有云。

⑩ Juno：提出 NFV 网络虚拟化概念；新增 Sahara 项目，支持用户大数据的集群部署；新增 LDAP 可集成 KeyStone 认证。

⑪ Kilo：Horizon 支持向导式创建虚拟机；Nova 部分标准化了 Conductor、Compute 与 Scheduler 的接口，为之后的接口分离做好准备；Glance 增加自动镜像转化格式功能。

⑫ Liberty：Neutron 提供更好的管理安全和带宽，更方便向 IPv6 迁移，LBaaS 已经成为生产化工具；Glance 基于镜像签名和校验，提升安全性；Swift 提高基本性能和可运维功能；Keystone 增加混合云的认证管理；引入容器管理的 Magnum 项目，通过与 OpenStack 现有的组件如 Nova、Ironic 与 Neutron 的绑定，Magnum 让容器技术的采用变得更加容易。

（2）OpenStack 项目架构

为了实现云计算的各项功能，OpenStack 实现了每个项目的既定目标，将存储、计算、监控和网络服务划分为几个项目来进行开发，每个项目也是对应的 OpenStack 中的一个或多个组件，图 1-6 所示为 OpenStack 的整体项目架构。

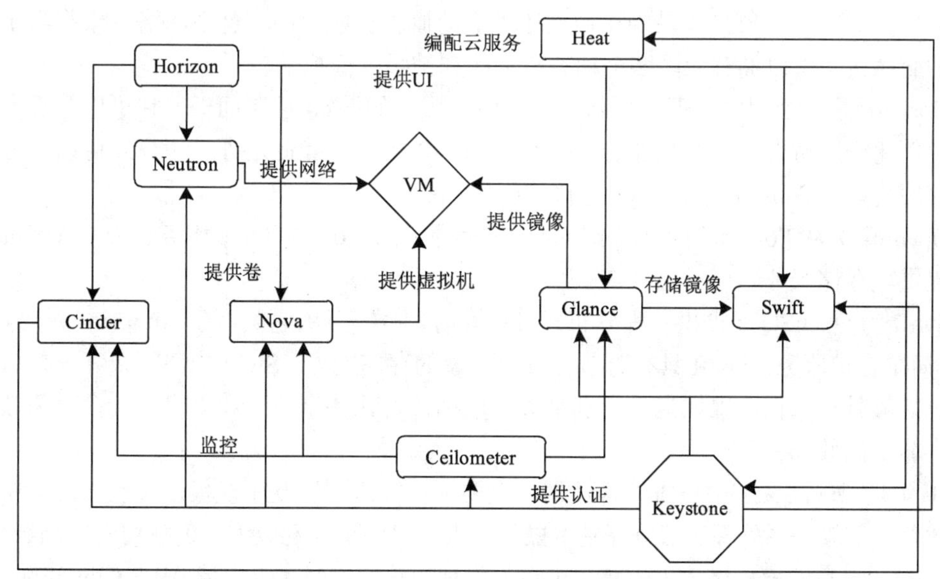

图 1-6　OpenStack 项目架构

OpenStack 各个组件之间的耦合是非常松的。其中，Keystone 是各个组件之间的通信核心，它依赖自身 REST（REpresentation State Trasfer）（基于 Identity API）对所有的 OpenStack 组件提供认证和访问策略服务，每个组件都需要向 Keystone 进行注册，主要目的是对云平台各个组件进行认证与授权，对云平台用户进行管理，注册完成后才能获取相对应的组件通信的地址，其中包括通信的端口和 IP 地址，然后实现组件之间和内部子服务之间的通信。

Nova 是 OpenStack 云计算的弹性控制器。OpenStack 云实例生命期所需的各种动作都将由 Nova 进行处理和支撑，这就意味着 Nova 以管理平台的身份登场，负责管理整个云的计算资源、网络、授权及测度。虽然 Nova 本身并不提供任何虚拟能力，但是它可使用 Libvirt API 与虚拟机的宿主机进行交互。Nova 通过自身的 API 对外提供处理接口，而且这些接口与 Amazon 的 Web 服务接口是兼容的。

OpenStack 镜像服务器是一套虚拟机镜像的发现、注册和检索系统，我们可以将镜像存储到以下任意一种存储中。

- 本地文件系统（默认）。
- S3 直接存储。
- S3 对象存储（作为 S3 访问的中间渠道）。
- OpenStack 对象存储。

镜像服务 Glance 主要提供两个服务。

- Glance API：主要负责接收响应镜像管理命令的 RESTFUL 请求，分析消息请求信息并分发其所带的命令（如新增、删除、更新等）。默认绑定端口是 9292。
- Glance Registry：主要负责接收响应镜像元数据命令的 RESTFUL 请求。分析消息请求信息并分发其所带的命令（如获取元数据、更新元数据等）。默认绑定端口是 9191。

Neutron 网络的目的是为 OpenStack 云更灵活地划分物理网络，在多租户环境下提供给每个租户独立的网络环境。另外，Neutron 提供 API 来实现这种目标。Neutron 中的用户可以创建自己的网络对象，如果要和物理环境下的概念映射的话，这个网络对象相当于一个巨大的交换机，可以拥有无限多个动态可创建和销毁的虚拟端口。

Horizon 是一个用以管理、控制 OpenStack 服务的 Web 控制面板，它可以管理实例、镜像，创建密匙对，对实例添加卷，操作 Swift 容器等。除此之外，用户还可以在控制面板中使用终端（Console）或 VNC 直接访问实例。

OpenStack 从 Folsom 开始使用 Cinder 替换原来的 Nova Volume 服务，为 OpenStack 云平台提供块存储服务。

Swift 为 OpenStack 提供一种分布式、持续虚拟对象存储，它类似于 Amazon Web Service 的 S3 简单存储服务。Swift 具有跨节点百级对象的存储能力。Swift 内建冗余和失效备援管理，也能够处理归档和媒体流，特别是对大数据（吉比特字节）和大容量（多对象数量）的测度非常高效。

Heat 是 OpenStack 中负责编排计划的主要项目。它可以基于模板来实现云环境中资源的初始化，依赖关系处理、部署等基本操作，也可以解决自动收缩、负载均衡等高级特性。目前，Heat 自身的模板格式（HOT）正在不停地改进，同时，也支持 AWS CloudFormation 模板（CFN），HOT 的目标是在不远的将来可以完全地替代 CFN。

Ceilometer 是 OpenStack 中的一个子项目，它像一个漏斗一样，能把 OpenStack 内部发生的几乎所有的事件都收集起来，然后为计费和监控以及其他服务提供数据支撑。

2. IaaS 云服务商

根据 Gartner 2015 年云计算 IaaS 云服务提供商分析图（见图 1-7），可以看到 Amazon、Microsoft 仍然是行业的领导者。但是，国内最大的云服务商阿里云并没有出现在这张图上。

目前 Amazon 提供了虚拟服务器 EC2、托管 Hadoop 框架 EMR、虚拟私有云 VPC、负载均衡、弹性伸缩、云存储 S3、块存储 EBS、内容分发 CDN 和数据库等基本服务。阿里云对应也提供了云主机 ECS、块存储、专有网络 VPC、负载均衡、弹性伸缩、对象存储 OSS、块存储、内容分发 CDN、数据库和安全等服务。

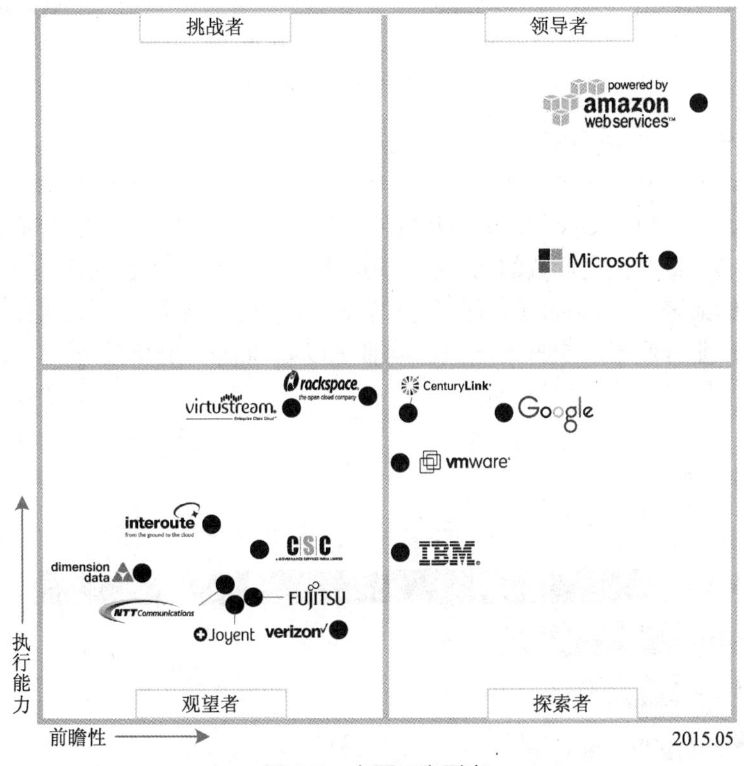

图 1-7　主要厂商列表

OpenStack 平台提供的服务包括云主机 Nova、网络 Neutron、对象存储 Swift、块存储 Cinder、大数据集群 Sahara 和数据库 Trova 等服务。

3. OpenStack 基金会

OpenStack 基金会（OpenStack Fundation）促进了 OpenStack 云操作系统在全球范围内的研发、分发和部署。目前，参与基金的个人成员来自 170 多个国家，超过 30 000 人。OpenStack 基金会目前得到了国内外企业的支持，如图 1-8 所示，其中白金会员有 AT&T、Canonical、HP、IBM、Intel、Rackspace、Red Hat、SUSE，金牌会员有 Cisco、Dell、EMC、Mirantis、Yahoo、华为等，其他支持公司包括希捷、Oracle、九州云、浪潮等。另外，国内外有大量支持和使用 OpenStack 的公司，如新浪、爱奇艺、360、小米、联想、携程、华胜天成、CETC55 等。根据《OpenStack 架构设计指导》(OpenStack Architecture Design Guide)，OpenStack 可应用于包括通用型、计算型、存储型、网络型、跨域型、混合型和大规模弹性扩展型的云平台场景。

图 1-8 OpenStack 主要基金会成员

4. OpenStack 项目调研

2015 年 10 月，OpenStack 社区发布了 OpenStack 用户报告（OpenStack User Survey），这份报告主要反映用户对 OpenStack 项目的使用状态和反馈的情况。报告显示，OpenStack 的技术日益成熟，目前接受调查的用户所使用的技术有 60%部署在生产环境中。

调研统计的部署 OpenStack 的行业分布情况如下，信息技术、科研、电信行业占据前 3 位，其中信息产业行业超过 58%的组织已经部署 OpenStack，其中接近一半用于生产系统，如图 1-9 所示。

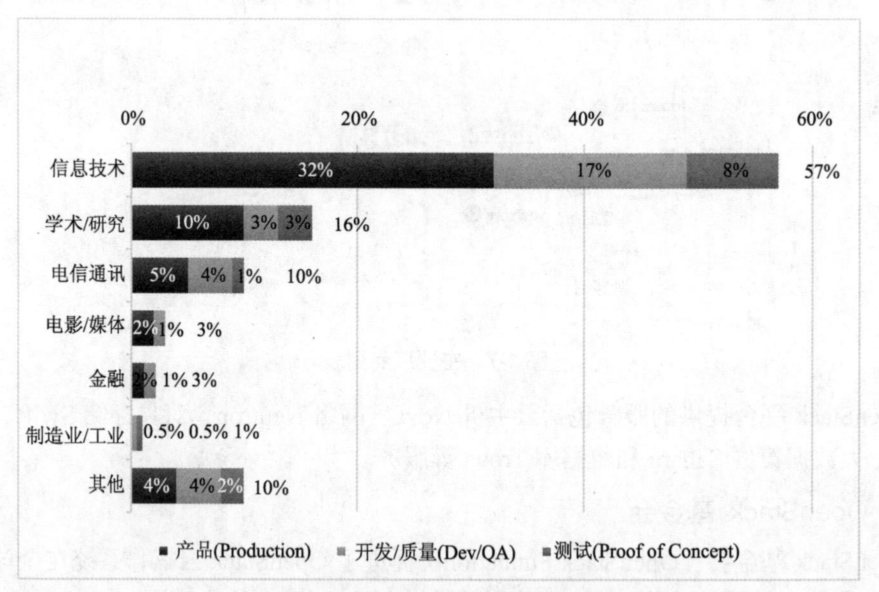

图 1-9 OpenStack 行业分布图

OpenStack 未来在新兴技术方面，包括在容器、网络功能虚拟化（NFV）、平台即服务（PaaS）方面的发展潜力巨大，如图 1-10 所示。

在 OpenStack 最终使用用户方面，报告也做了一定的调查，图 1-11 所示为终端用户的显示图，使用最多的为北美地区，占到整体的 44%，其次为亚洲地区，为 28%，欧洲以 22% 占据第 3；如果按照国家来排名，前 5 名分别为美国 39%，中国 8%，印度 7%，日本 6%，法国 4%。

在所有 OpenStack 生产环境中，图 1-12 所示的 Icehouse 以 39%的部署率占据第 1 位位置，第 2 位 Juno 为 35%，第 3 位 Kilo 为 30%，这些都说明新版本都在不断地受到用户的欢迎和支持。

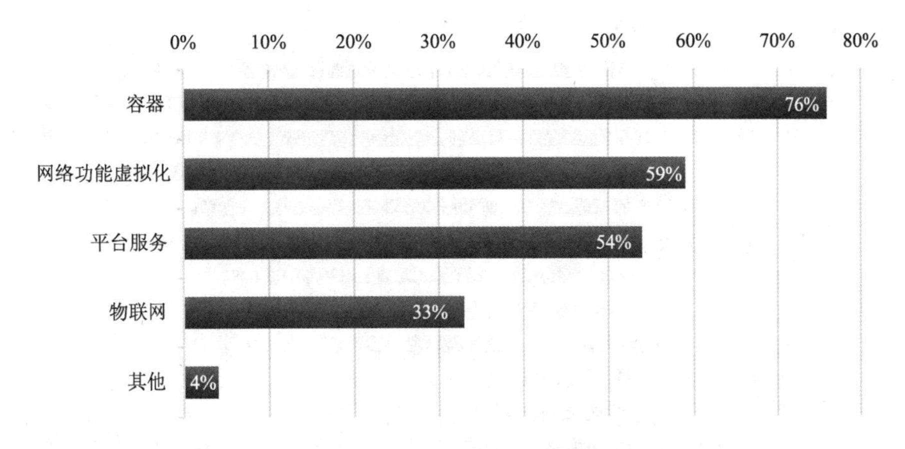

图 1-10　OpenStack 新兴技术关注热点

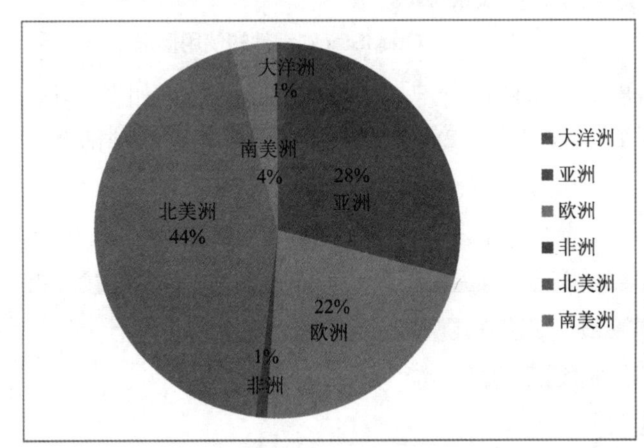

图 1-11　OpenStack 用户分布情况

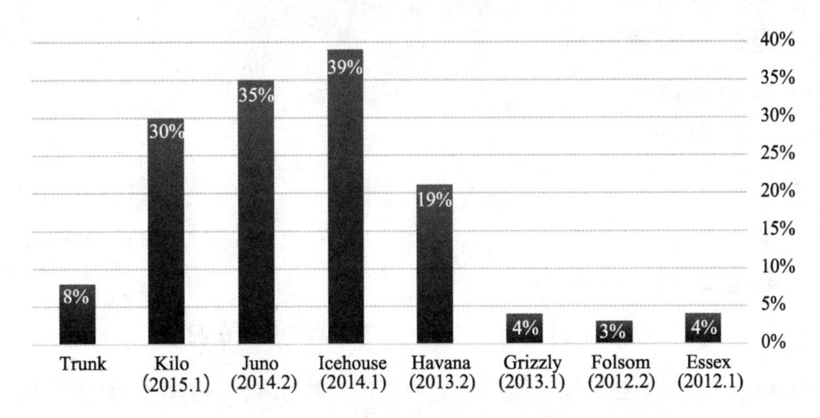

图 1-12　OpenStack 部署使用图

OpenStack 提供众多的组件给用户使用，但是什么组件最受用户欢迎和接受一直以来没有具体的表现，这份报告向我们指出了组件的使用情况。如图 1-13 所示，Nova、Keystone、Glance、Neutron、Cinder 仍然是 OpenStack 的核心组件，其使用率最高。

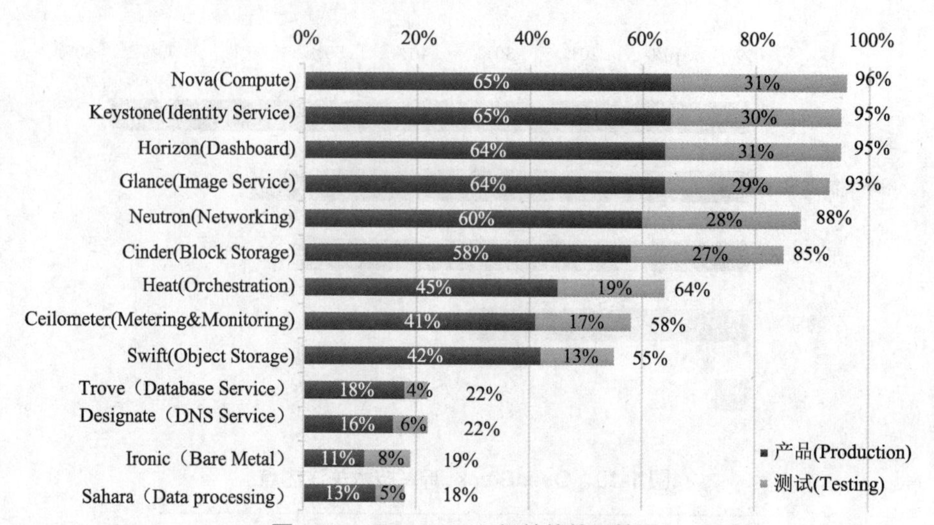

图 1-13　OpenStack 组件的使用情况

调研报告还指出了众多的流行部署工具的使用情况，如图 1-14 所示。问卷中 37%的生产环境是使用 Puppet 部署，28%为 Ansible 部署，这两种部署工具占据一半以上的使用量。

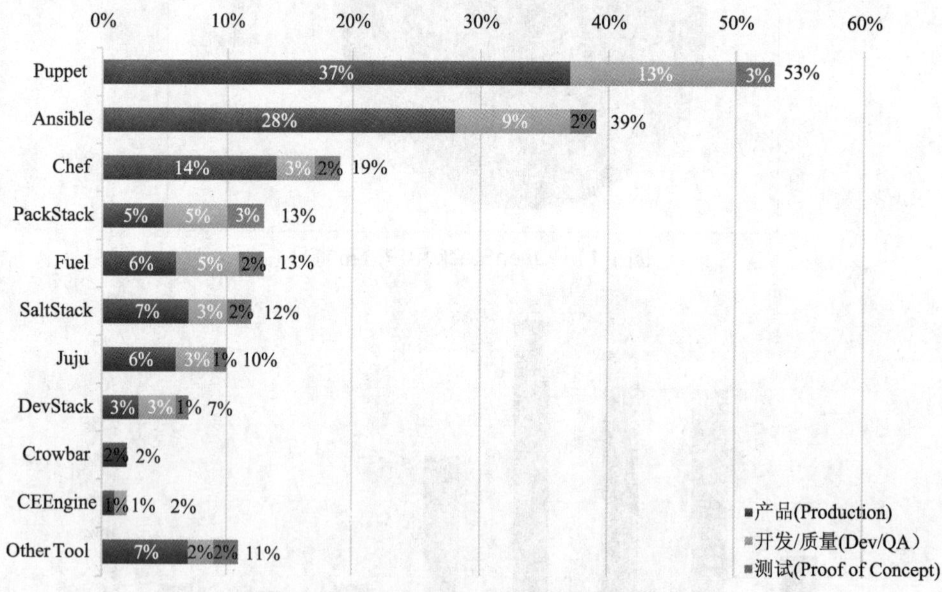

图 1-14　OpenStack 部署工具的使用情况

在 OpenStack 的虚拟化类型的选择上，调查报告也给我们指出了方向，如图 1-15 所示。57%的用户选择 KVM 作为其虚拟化管理程序（Hyperviosr），说明 KVM 发展的同时也促进了 OpenStack 的发展和进步。

提示

对于 OpenStack 的认识重点是要掌握其包含的 12 个核心项目，并对其整体项目架构有一个清晰的概念，从而为搭建开源的 OpenStack 云计算平台提供有力的技术基础。OpenStack 可应用于包括通用型、计算型、存储型、网络型、跨域型、混合型和大规模弹性扩展型的云平台场景。

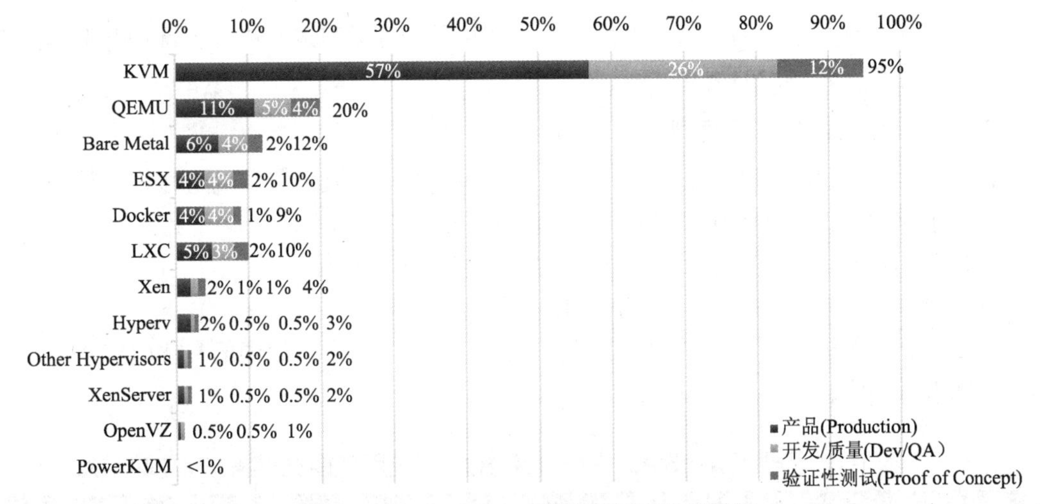

图 1-15　OpenStack 虚拟化类型的选择情况

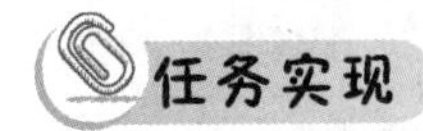
任务实现

1. OpenStack 的技术资源

可通过以下几种途经查找 OpenStack 相关的技术资源。

（1）OpenStack 官网

- http://www.OpenStack.org/。
- https://wiki.OpenStack.org/wiki/Main_Page。
- OpenStack 中文文档：http://docs.OpenStack.org/zh_CN/。

（2）OpenStack - 开源中国社区

- http://www.oschina.net/question/tag/OpenStack。
- 开源中国社区：http://www.oschina.net/。

（3）中外文的相关资源

在搜索引擎上查找 OpenStack 核心模块对应的网站或 Blog。

（4）专业的 IT 公司的服务支持

专业公司的服务起到了中国 OpenStack 服务中心的作用，可以更方便地获得技术支持。

① 南京第五十五所技术开发有限公司。

南京第五十五所技术开发有限公司提供有关 OpenStack 云计算平台的服务。

- 提供企业与教育行业的物联网技术解决方案。
- 提供开源的 OpenStack 云计算平台搭建服务与职业技能竞赛。
- 提供基于云计算的在线课程系统、云计算在线实训系统、云在线考试分析系统。

② 华胜天成公司。

华胜天成提供有关的技术服务。

- 为 OpenStack 的研究者、开发者和使用者提供丰富的线上及现场专业支持服务和咨询服务，消除客户使用开源软件的后顾之忧。
- 提高国内云计算从业人员数量和素质，普及开源软件精神与技术。

- 打破云计算建设的垄断，大大降低云计算平台建设与运营成本，推动并保障国内云计算平台建设的蓬勃发展。

2. OpenStack 的项目案例

以下基于 OpenStack 的 2016 年江苏省云计算技术与应用技能竞赛的考核知识结构和相关内容来了解 OpenStack 项目的实施环节与内容要求。

案例名称：2016 年江苏省云计算技术与应用技能竞赛项目。

项目要求：基于 OpenStack 的云计算服务平台的搭建。

项目的具体考核环节与考核知识点及技能点如表 1-1 所示。表中的有些内容并不属于本课程的教学内容，可作为了解内容。

表 1-1　基于 OpenStack 的云计算技术与应用技能竞赛考核要求

序号	考核环节	考核知识点和技能点
1	云计算基础架构平台（IaaS）	按照试题系统网络架构要求，检查网络设备和服务器设备连线、配置是否正确
		Linux 操作系统准备和系统配置，通过系统的配置文件查看正确性
		基本服务 SELinux、NTP、MySQL 和 QPID 的安装、配置和使用。通过配置文件或验证命令查看正确性
		IaaS 平台安全统一框架服务的安装、配置和管理，并对脚本、配置和运行错误进行排查和改正。通过配置文件或验证命令查看正确性
		IaaS 平台镜像服务的安装、配置和管理，并对脚本、配置和运行错误进行排查和改正。通过配置文件或验证命令查看正确性
		IaaS 平台计算服务的安装、配置和管理，并对脚本、配置和运行错误进行排查和改正。通过配置文件或验证命令查看正确性
		IaaS 平台网络服务的安装、配置和管理，并对脚本、配置和运行错误进行排查和改正。通过配置文件或验证命令查看正确性
		IaaS 云平台软件定义网络（SDN）控制器、网络交换节点的安装、配置和管理，并对脚本、配置和运行错误进行排查和改正。通过配置文件或验证命令查看正确性
		IaaS 平台控制面板的安装、配置和管理，并对脚本、配置和运行错误进行排查和改正。通过配置文件或验证命令查看正确性
		IaaS 平台存储服务的安装、配置和管理，并对脚本、配置和运行错误进行排查和改正。通过配置文件或验证命令查看正确性
		IaaS 平台模板和监控服务的安装、配置和管理，并对脚本、配置和运行错误进行排查和改正。通过配置文件或验证命令查看正确性
		IaaS 大数据整合服务的安装、配置和管理，并对脚本、配置和运行错误进行排查和改正。通过配置文件或验证命令查看正确性

续表

序号	考核环节	考核知识点和技能点
2	云计算开发服务平台（PaaS）	部署 PaaS 平台控制节点（Controller）和资源节点（Node）云主机，部署成功后，修改操作系统参数和服务。通过配置文件或验证命令查看正确性
		在资源节点（Node）云主机上安装 PaaS 资源模块，通过配置文件或验证命令查看正确性
		在控制节点（Controller）云主机上安装 PaaS 管理模块，通过配置文件或验证命令查看正确性
		配置和联通控制节点（Controller）和资源节点（Node），完成 PaaS 平台的安装。通过配置文件或验证命令查看正确性
		注册和安装 PaaS 平台应用套件，完成 PaaS 平台的中间件、软件和开发环境的资源搭建。通过访问 PaaS 平台进行验证，查看正确性
3	云计算软件服务平台（SaaS）	部署开发测试环境，构建开发测试云服务（Python、Python Django、Ruby、Node.js、Java JBoss、PHP、MySQL、MongoDB）。通过配置文件或验证命令查看正确性
		部署企业应用，构建企业云应用服务（客户关系管理（CRM）、博客系统（Blog）、企业资源规划计划（ERP）、内容管理系统（CMS）等）。通过配置文件或验证命令查看正确性
		整合 VMware 企业级虚拟化，构建云桌面服务（VDI）。通过配置文件或验证命令查看正确性
		使用、管理和监控云桌面服务（VDI），通过配置文件或验证命令查看正确性
4	大数据平台部署	进行 Hadoop HDFS 和 Map-Reduce 的安装和配置，制作 Sahara 服务的 Hadoop 节点集群镜像
		注册 Sahara 的 Hadoop 节点集群镜像，配置节点模板，配置集群模板，通过集群模板部署 Hadoop 集群
		数据仓库 Hive 的部署、配置和应用，部署成功后使用 Hive 进行数据仓库的增、删、查、改和管理操作
		分布式数据库 HBase 部署、配置和应用，部署成功后使用 HBase 进行分布式数据库的增、删、查、改和管理操作
		内存计算框架 Spark 的部署、配置和应用，部署成功后运行 Spark 分析案例
		机器学习和数据挖掘框架 Mahout 的部署、配置和应用。部署成功后运行 Mahout 案例
5	云应用开发	搭建云应用 Web 系统开发环境，进行云应用 Web 程序开发
		基于 Web 开发环境，开发云应用网盘 Web 系统，实现电子文件的列表管理、预览、上传、下载、删除、复制、移动、重命名和管理等功能

续表

序号	考核环节	考核知识点和技能点
5	云应用开发	发布云应用网盘 Web 系统，提交发布后的 Web 应用程序。压缩云应用网盘 Web 系统源代码，提交压缩后的 zip 文件
		搭建 Android 移动 App 开发环境，进行云应用 App 客户端开发
		基于 Android 4.4 App 开发环境，开发云应用网盘 App 客户端，实现电子文件的列表管理、预览、上传、下载、删除、复制、移动、重命名和管理等功能
		编译云应用网盘 App 客户端，提交编译后的 apk 程序。压缩云应用网盘 App 客户端源代码，提交压缩后的 zip 文件
		搭建大数据分析和挖掘开发环境，进行数据分析应用开发
		基于大数据分析和挖掘开发环境，使用案例数据进行 Map-Reduce 大数据分析应用开发
		基于大数据分析和挖掘开发环境，使用案例数据进行 Spark 大数据分析应用开发
		对 Map-Reduce 和 Spark 的分析结果进行对比、分析
6	文档及职业素养	工程文档编写，编写系统结构、功能需求、功能测试或项目实施报告
		比赛现场符合企业“5S”（即整理、整顿、清扫、清洁和素养）原则
		团队分工明确合理、操作规范、文明竞赛

提示

对于一个系统化的复杂开源云计算平台的搭建任务，除了要掌握基础概念外，还应该针对在业界比较有影响力的开源平台进行深入地研究，如 OpenStack 的性能、架构及应用的广泛性。还必须充分地获取有关的技术资源，能够借鉴成功的项目案例，少走弯路。还需要开发人员具有独立的研发能力和项目开发经验，用好开源技术的优势。

项目二 环境设计和系统准备

本项目主要帮助读者掌握搭建 OpenStack 云计算平台的环境设计及系统准备，包括硬件基本需求，OpenStack 云计算平台搭建所需的软件包，部署一个实际的 OpenStack 云计算平台拓扑结构，并在这个环境下进行系统安装基础工作。

学习目标

- 掌握构建云计算平台的系统拓扑结构。
- 掌握系统拓扑结构下的网络配置。
- 掌握正确搭建云计算平台的安装基础工作。

任务一 云计算平台的系统架构

任务要求

小李基本掌握了云计算平台搭建的基础知识，接下来需要对公司的应用需求进行调研，在此基础上要进行公司云计算平台的系统环境设计和系统搭建的基础安装工作，为此，小李当前要完成的任务如下。

- 公司云平台应用的需求分析。
- 公司云平台系统环境架构设计。

相关知识

1. 项目需求分析

（1）基本概念

需求分析是指理解用户需求，与用户的功能需求达成一致，并估计项目风险和评估项目代价，最终形成开发计划的一个复杂过程。在这个过程中，用户是处在主导地位的，需求分析工程师和项目经理要负责整理用户需求，为之后的项目设计打下基础。

从广义上理解，需求分析包括需求的获取、分析、规格说明、变更、验证、管理等一系列需求工程。狭义上理解，需求分析指需求的分析、定义过程。

需求分析阶段结束后应该得到相应的需求分析报告。

（2）分析内容

需要分析的内容可以包含公司应用需求、技术资金投入与生产效益、行业技术发展趋势、国家政策支持等。

（3）分析过程

需求分析阶段的工作可以分为 4 个方面：问题识别、分析与综合、制订规格说明、评审。

（4）分析方法

需求分析的方法有很多，如原型化方法、结构化方法和动态分析法等。

2. 系统架构设计

一个项目的系统架构设计一般是由系统架构设计师来负责完成的。对于系统架构设计师来说，其主要职责有如下 4 条。

（1）确认需求

在项目开发过程中，架构师是在需求规格说明书完成后介入的，需求规格说明书必须得到架构师的认可。架构师需要和分析人员反复交流，以保证自己完整并准确地理解用户需求。

（2）系统分解

依据用户需求，架构师将系统整体分解为更小的子系统和组件，从而形成不同的逻辑层或服务。随后，架构师会确定各层的接口、层与层相互之间的关系。架构师不仅要对整个系统分层，进行“纵向”分解，还要对同一逻辑层分块，进行“横向”分解。

（3）技术选型

通过对系统的一系列的分解，架构师最终形成项目的整体架构。技术选择主要取决于项目架构。

架构师对产品和技术的选型仅仅限于评估，没有决定权，最终的决定权归项目经理。架构师提出的技术方案为项目经理提供了重要的参考信息，项目经理会对项目预算、人力资源和时间进度等实际情况进行权衡，最终进行确认。

（4）制定技术规格说明

在项目开发过程中，架构师是技术权威。他需要协调所有的开发人员，与开发人员一直保持沟通，始终保证开发者依照他的架构意图去实现各项功能。

架构师不仅要保持与开发者的沟通，也需要与项目经理、需求分析员，甚至最终用户保持沟通。所以，对于架构师来讲，不仅有技术方面的要求，还有人际交流方面的要求。

3. 环境说明

① 若教学环境有足够的可供学生使用的服务器，则每组分配两台服务器进行练习。

② 若教学环境没有服务器，可使用 PC 代替，每组分配两台 PC 进行练习（每台 PC 都需支持 CPU 虚拟化，配置双网卡，最低 4GB 内存，最低 100GB 硬盘）。

③ 若教学环境可供学生使用的服务器数量不够，可将所有服务器组建成一个云平台，每组学生分配两台虚拟机进行练习（每台虚拟机配置双网卡，最低 4GB 内存，最低 100GB 硬盘）。

提示

有了项目的需求分析和系统架构设计，才能针对具体的情况按照项目系统解决方案去实施项目任务。通过实际项目的锻炼是最好的学习方式。

任务实现

1. 公司对云平台应用的需求

经过调研分析，公司的基本情况如下。

（1）公司的基本组织结构

内部有 100 名员工，其中 50 名在项目研发部（研发环境），45 名在业务部（办公环境），5 名在 IT 工程部（运维环境）。

（2）应用需求情况

按员工的办公情况不同，分别使用 CentOS 6.5、Ubuntu、Windows 7 和 Windows Server 镜像办公；根据云存储特点，将镜像资源云硬盘存储于 Swift 内部，提升镜像的安全性；编写批量模板文件，可以在短期内快速部署集群；构建内部块存储和卷存储，实现实例扩容和公司内部资源存储；根据企业员工的构成比例构建 4 种办公网络和 4 个租户组，保证单位内部资源隔离和资料安全；使用监控系统可以查看平台运行情况，保证系统的正常稳定运行，以及监测硬件平台的稳定。

（3）服务需求

- 构建 3 个用户租户，100 个用户，管理人员拥有管理员权限，其余人员拥有普通用户权限。
- 构建 4 种不同类别镜像，镜像后端存储为 Swift，满足企业正常的办公需求。
- 构建云硬盘服务，云硬盘后端存储为 Swift，保证资源的安全。
- 创建 3 个用户租户网络，满足正常的办公需求。
- 编写模板文件，定制虚拟机特性。
- 运行监控系统，查看系统运行情况。

2. 云平台系统架构设计

按照既定的项目目标，接下来将围绕这个目标开始一步一步构建云计算平台，以满足日常的企业办公、生产和研发需求。基于以上要求，依据 OpenStack 架构指南，构建一个通用性云平台。遵循 IaaS（Infrastructure as a Service）模式，基于简单的需求为用户寻求最合适的平台。

通用性平台为最基本、最简单的平台，适合概念验证、小型实验，也可以基于通用性平台随意扩展计算资源和存储资源。整套平台环境的网络拓扑结构如图 2-1 所示。

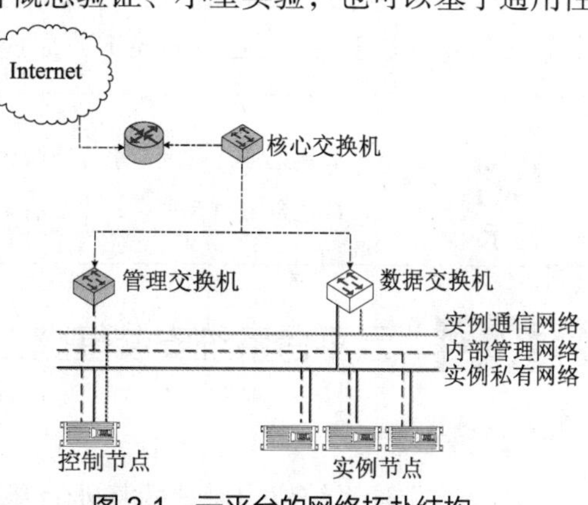

图 2-1　云平台的网络拓扑结构

（1）拓扑结构说明

在云平台的网络拓扑结构中，采用两种节点服务器构建云计算平台，其中一种为控制节点服务器，另一种为实例节点（即计算节点，以下相同）服务器。

按照网络分离和功能化要求，依次构建了 4 种网络，分别为实例通信网络、内部存储网络、内部管理网络和实例私有网络。同时，考虑到服务器只有两个网口的实际情况，采取结合 OpenvSwitch 虚拟交换机功能的方法虚拟生成 3 个网口，对应为 br-ex、br-mgmt 和 br-prv，分别作为实例通信网络、内部管理网络和内部私有网络。将服务器流量转化为两个交换机来进行数据流量分流，交换机 1 对应内部管理网络流量，交换机 2 对应实例私有网络和实例通信网络流量。

（2）系统架构设计

每个节点模块的安装服务也是根据系统拓扑结构来进行确定的。根据前期系统的部署，将平台服务做了拆分，如节点部署服务如图 2-2 所示。

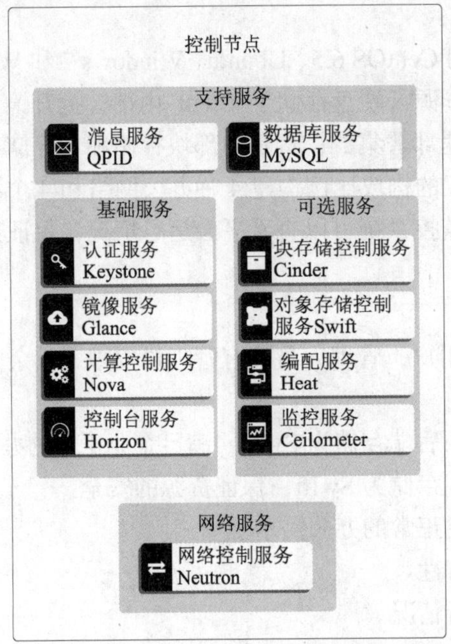

图 2-2　节点部署服务示意图

① 控制节点安装如下服务：消息服务（QPID）、数据库服务（MySQL）、认证服务（Keystone）、镜像服务（Glance）、计算控制服务（Nova）、网络控制服务（Neutron）、控制台服务（Horizon）、块存储控制服务（Cinder）、对象存储控制服务（Swift）、编配服务（Heat）和监控服务（Ceilometer），完成云平台控制端的安装工作。

② 实例节点安装如下服务：计算服务（Nova Compute）、网络控制服务（Neutron）和监控服务（Ceilometer Compute），完成实例节点的安装。

提示

必须先将公司对云平台应用的需求分析清楚，才能针对具体的情况制定出合理的解决方案，搭建合理的云平台，才能将平台实际地应用起来。

任务二　云平台系统安装基础工作

任务要求

小李在上述系统架构设计的基础上开展下一步的工作，他准备好 OpenStack 搭建云计算平台项目所需的软件资源包，按云平台的网络拓扑结构图进行设备准备与网络连接，完成云平台系统安装基础工作。

- 准备 OpenStack 搭建云计算平台项目所需的软件资源包。
- 确定各节点的名称。
- 配置各节点的 IP 网络地址。
- 按要求安装各节点的操作系统。
- 配置系统环境变量。
- 在控制节点、网络节点、实例节点和存储节点分别运行脚本，完成各节点的配置安装。
- 验证安装基础工作。

相关知识

1. 节点主机名及 IP 地址规划

根据图 2-1 所示的拓扑结构图，本次部署的各节点 IP 地址分配如表 2-1 所示，同时，各节点的主机名称也规划在表中。

表 2-1　各节点主机名和 IP 地址规划列表

节点 / 主机	主机名	IP 规划		
		实例通信	内部管理	内部私有
控制节点	controller	172.24.3.10	172.24.2.10	172.24.4-6.0/24
实例节点	compute	172.24.3.20	172.24.2.20	172.24.4-6.0/24

2. 各节点的安装要求

（1）主机要求

主机为双网卡服务器或者 PC，内存为 4GB 以上，处理器 2 核以上。

（2）系统要求

操作系统为 CentOS 6.5_x64bit。

（3）节点

节点主要指部署云平台的物理节点服务器，下面来说明实验环境节点的要求和作用。

控制节点：存放系统数据库、中间件服务，实际为云平台系统的大脑和控制中心。

实例节点：存放虚拟机的服务器，支持处理器虚拟化功能，运行虚拟机管理程序（QEMU 或 KVM），管理虚拟机主机，同时为外部用户提供存储服务，为内部实例提供块存储服务。

3. 与 Linux 相关的操作知识

（1）Linux 系统安装

OpenStack 云计算平台的搭建过程中需要重点掌握的一些基础操作知识如下。

① Linux 的版本：Linux 的版本可分为内核版本与发行版本。

● 内核版本

内核版本是 Linux 任何版本的核心，内核版本号的组成形式为 X.Y.Z。

X：主版本号，通常在一段时间内比较稳定。

Y：次版本号，偶数代表正式版本，可公开发行，奇数代表测试版本，不太稳定。

Z：修改号，表示第几次修改，如 2.6.20。

● 发行版本

发行版本是将 Linux 内核与应用软件打包发行的版本，主流的 Linux 发行版本有 Red Hat Enterprise Linux 6.0、CentOS 6.5、Ubuntu、Debian GNU/Linux、openSUSE 和红旗 Linux 等。

这里 OpenStack 云计算平台的搭建使用 CentOS 6.5_x64bit 版本。

② 云服务器的安装要求为，一般服务器中包含处理器模块、存储模块、网络模块、电源和风扇等设备。云服务器关注的是高性能吞吐量计算能力，是在一段时间内的工作量总和。因此，云服务器在架构上和传统的服务器有着很大的区别，具有庞大的数据输入量或海量的工作集。

● 硬件配置

如果是云计算的应用需要，一般需购买带有双网卡（或多网卡）的高性能服务器。

如果用 OpenStack 搭建云计算平台只是做一些实验工作，也可以用普通高配置计算机加两块网卡。

● 硬盘分区

对服务器进行初始安装时可在本地用光盘或 U 盘进行系统安装，对控制节点的硬盘分区没有特殊的要求，但是，对计算节点（实例节点）一般可预留两个空分区，如图 2-3 所示。

在图 2-3 中，sda2 和 sda3 是两个预留的空分区，这里预留分区的大小只是做一个演示，不能作为实际分区的大小，并且要记住这个分区的设备名称，也可以到/dev 目录下查看设备名称 sda2、sda3，这在以后的存储节点磁盘分区时会用到。

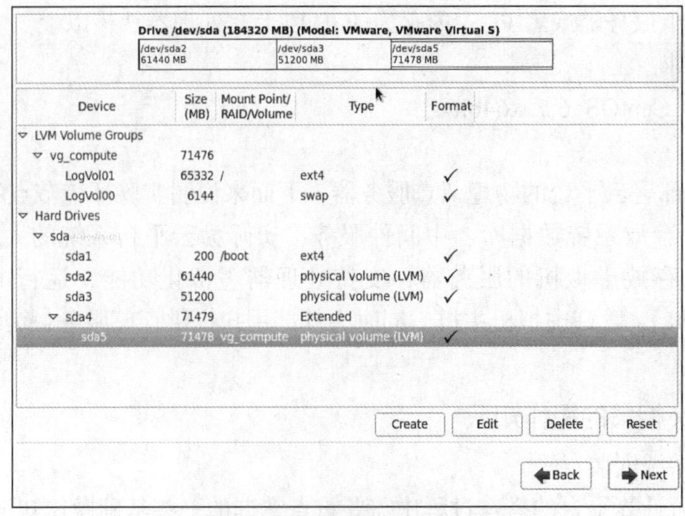

图 2-3 实例节点硬盘分区图

● 最小化安装

安装时需要注意以下几个方面：在进行安装时请选择英文界面；填写主机名称（hostname）；单击“Configure Network”进行 IP 地址的设置；取消选择“System clock uses UTC”项；选择“Create Custom Layout”进行系统分区；选择安装系统时建议选择最小化（Minimal）安装，如图 2-4 所示。

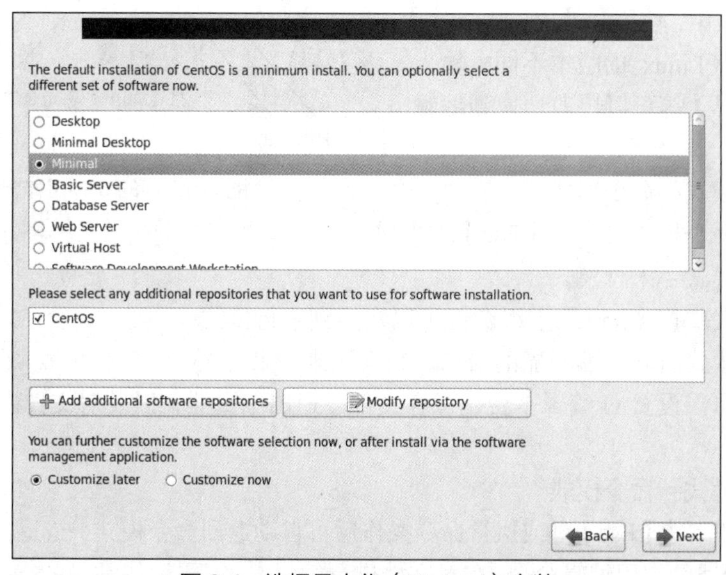

图 2-4　选择最小化（Minimal）安装

即使是最小化安装中也会默认安装 SSH 服务，这样就可以进行远程登录管理操作，完成云计算平台的搭建。

（2）系统配置文件

常用的系统配置文件见表 2-2。

表 2-2　常用的系统配置文件

序号	配置文件	所在子目录	功能
1	hosts	/etc	主机名与 IP 地址的映射关系
2	network	/etc/sysconfig/	主机名称
3	ifcfg-eth0	/etc/sysconfig/network-scripts/	网卡 0 的 IP 地址
4	config	/etc/selinux/	SELinux 的配置
5	iptales	/etc/sysconfig/	配置防火墙规则

（3）vi 的使用

vi 是一个功能强大的全屏幕文本编辑工具，一直以来都作为类 UNIX 操作系统的默认文本编辑器。vim 是 vi 编辑器的增强版本，在 vi 编辑器的基础上扩展了很多实用的功能，但是习惯上还是将 vim 称作 vi。

① vi 的工作模式：vi 编辑器有 3 种工作模式，如图 2-5 所示。

在使用 vi 编辑文本时，只有时刻清楚自己处在图 2-5 中的哪种模式下，才能知道下一步的操作用什么命令来完成。

当输入“vi filename”命令后，即进入 vi 的命令模式，如图 2-5 所示。vi 的 3 种模式间的切换可通过以下按键进行。

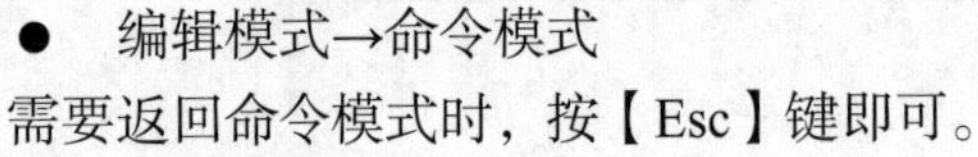

图 2-5　vi 的 3 种工作模式

- 命令模式→编辑模式

在命令模式中，使用【a】、【i】、【o】或【Insert】键（Linux 的版本不同，输入的命令键有差异）等可以快速切换到编辑模式。

- 编辑模式→命令模式

需要返回命令模式时，按【Esc】键即可。

- 命令模式→末行模式

在命令模式中按【Shift+:】组合键可以切换到末行模式。

处于末行模式时，vi 编辑器的最后一行会出现“:”提示符。

该模式中可以设置 vi 编辑平台、保存文件、退出编辑器，以及对文件内容进行查找、替换等操作。

- 末行模式→命令模式

在末行模式下按【Esc】键或执行命令操作后可自动返回命令模式，或退出 vi 操作命令。

② vi 编辑器的常用编辑命令：在对文档进行编辑时，需要记住常用的操作功能键命令，通过不断实践逐步熟练掌握。常用的编辑功能键盘操作列表，如表 2-3 所示。

表 2-3　常用的编辑功能键盘操作列表

工作模式	键盘操作	完成的操作功能
命令模式	【x】或【Del】键	删除光标处的单个字符
	dd	删除当前光标所在行
	yy	复制当前行整行的内容到剪贴板
	*yy	*号表示行数，复制从光标处开始的*行内容到剪贴板
	【p】键	可将内容粘贴到光标位置处之后
	/ str	从当前光标处开始向后查找字符串 str
	? str	可以向前查找字符串 str
	【n】或【N】键	分别为在查找的结果中向后或向前继续查找
	【u】键	取消最近一次的操作，并恢复操作结果
	【U】键	用于取消对当前行的所有编辑
	gg	将光标移至文档的首部
	【Shift+g】组合键	将光标移至文档的尾部
	ZZ	保存当前的文件内容并退出 vi

续表

工作模式	键盘操作	完成的操作功能
末行模式	:w	保存文件
	:wq	保存并退出 vi
	:w 文件名	另存为其他文件名
	:q	没有修改文档内容，退出 vi
	:q!	不保存文档修改，强制退出 vi
	:set nu	显示文档行号
	:set nonu	取消显示行号
	:d:	删除当前行
	:nd:	删除从当前开始的 n 行
输入模式	【↑】、【↓】、【→】、【←】键	光标移动方向
	【PageDown】键	向下翻页
	【PageUp】键	向上翻页
	【Home】键	光标快速跳转到本行的行首
	【End】键	光标快速跳转到本行的行尾
	【Del】键	删除光标所在位置的字符
	【Backspace】键	删除光标所在位置前面的字符

（4）“yum”命令的使用

① yum 是什么?

yum 即 yellow dog updater, modified，主要功能是更方便地添加或删除更新 RPM 包，它能自动解决包的依赖性问题，便于管理大量系统的更新问题。

另外，APT（Advanced Packaging Tool）是一款强大的包管理工具，也可以称作机制，在 Debian 及其衍生版本的 GNU/Linux 中使用。为什么 CentOS 要使用 yum 而不用 APT?最简单的原因是 CentOS 自带 yum。

② yum 特点

- 可以同时配置多个资源库(Repository)。
- 简洁的配置文件(/etc/yum.conf，/etc/yum.repos.d 下的文件)。
- 自动解决增加或删除 RPM 包时遇到的依赖性问题。
- 使用方便。
- 保持与 RPM 数据库的一致性。

③ yum 安装

CentOS 自带 yum 软件包，在最小化安装时已经默认安装了 yum，也可用下面的命令安装，但在具体应用时需修改源的路径。

```
# rpm -ivh yum-3.2.29-40.el6.centos.noarch.rpm
```

CentOS 采用的二进制包是 RPM，CentOS 的 yum 使用方法和 APT 有几分相似，安装了 yum 后，许多包的安装就方便多了。在 OpenStack 云计算平台的搭建中需要改变软件包的文件位置，因此还需要进行相应的修改，后面的项目任务会进行相应的练习，但是，yum 是必须安装的。

④ CentOS 的 yum 配置。

- 加快 yum 下载速度。

```
# yum -y install yum-plugin-fastestmirror-1.1.30-14.el6.noarch.rpm
```

- yum 源配置文件的位置为/etc/yum.repos.d/*.repo。

⑤ 将 yum 源设为 FTP 网站，在/etc/yum.repos.d/下创建 centos65.repo 源文件。

```
# vi centos65.repo
```

配置内容如下。

```
[centos6.5]
name=CentOS6-5
baseurl=ftp://192.168.2.10/centos6.5
gpgcheck=0
enabled=1
```

做好配置后，在需要用“yum”命令进行安装时，就会到 FTP 服务器上指定的路径中读取软件安装包并进行安装。

⑥“yum”命令简介

当第一次使用 yum 或 yum 资源库有更新时，yum 会自动下载所有所需的 headers，并放置于/var/cache/yum 目录下，所需时间可能较长。

```
# yum check-update                //检查可更新的 RPM 包
# yum clean packages              //清除暂存中的 RPM 包文件
# yum list                        //列出资源库中特定的可以安装的 RPM 包
                                  //或更新已经安装的 RPM 包
# yum install xmms-mp3            //安装 RPM 包，如 xmms-mp3
# yum remove licq                 //删除 RPM 包，包括与该包有依赖性的包
```

（5）Linux 脚本命令的运行

为了简化云平台的安装过程，将搭建过程归纳为几大步骤，每一个步骤都通过 Linux 的命令脚本的执行完成相关的任务，实训过程中提供了相应的安装脚本。因此，对 Linux 的命令脚本的编程和执行知识有所了解，通过分析脚本，可以更清楚地知道安装过程及内部完成的任务，对云平台的搭建有很好的理解和掌握效果。

（6）FTP 服务器

在 OpenStack 云计算平台的搭建过程中需要安装系统的软件包，这些软件包可以直接复制到本地服务器上，然后建立本地源 yum，用“yum”命令进行软件包的安装操作。但是，有时不必将大量软件包复制到服务器上，而是在 Linux 客户端用其自带的 FTP 服务器

或在 Windows 客户端用第三方 FTP 服务软件（如 Serv-U、3CDamon）建立 FTP 服务器，在服务器上建立源 yum 的安装配置。这样，安装软件包也就更加方便。

提示

要根据所搭建的云平台系统网络拓扑结构图来规划各节点的 IP 地址、各节点的主机名称，并根据节点的硬件配置情况合理地调配节点的功能任务，对实训提供的安装包、安装脚本要预先有一个了解。

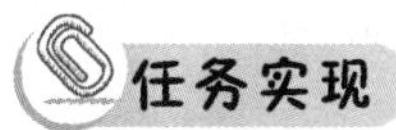

1. 云平台基础部署工作

（1）软件包资源和安装脚本

用 OpenStack 搭建云计算平台项目所需的软件包资源和安装脚本介绍，如表 2-4 所示。

表 2-4 用 OpenStack 搭建云计算平台项目所需的软件包资源和安装脚本

编号	软件包资源或安装脚本	作用
1	CentOS-6.5-x86_64-bin_DVD.iso	操作系统镜像
2	XianDian-IaaS-v1.4.iso	OpenStack 软件包
3	Xiandian_Pre.sh	环境初始配置脚本
4	Xiandian_Install_Controller_Node.sh	控制节点安装脚本
5	Xiandian_Install_Compute_Node.sh	实例节点安装脚本

（2）各节点安装操作系统

安装最小化 CentOS 6.5_x64 桌面操作系统，配置主机名，将提供的压缩包导入到操作系统内。

（3）配置主机名

配置控制节点主机名为“controller”，配置实例节点主机名为“compute”。

在控制节点修改配置文件/etc/sysconfig/network，内容如下。

```
NETWORKING=yes
HOSTNAME=controller
```

配置完成，通过“hostname”命令进行验证。

```
# hostname  controller
# hostname
controller
```

在实例节点修改配置文件/etc/sysconfig/network，内容如下。

```
NETWORKING=yes
HOSTNAME=compute
```

配置完成，通过“hostname”命令进行验证。

```
# hostname  compute
# hostname
compute
```

（4）配置域名解析

在全部节点的/etc/hosts 文件中添加域名解析。

在控制节点修改配置文件/etc/hosts，内容如下。

```
172.24.2.10     controller
172.24.2.20     compute
```

在实例节点修改配置文件/etc/hosts，内容如下。

```
172.24.2.10     controller
172.24.2.20     compute
```

（5）配置环境

配置防火墙规则，内容如下。

```
# iptables -F   //清除所有 chains 链（INPUT/OUTPUT/FORWARD）中所有的 rule 规则
# iptables -Z   //清空所有 chains 链（INPUT/OUTPUT/FORWARD）中的包及字节计数器
# iptables -X   //清除用户自定义的 chains 链（INPUT/OUTPUT/FORWARD）中的 rule 规则
# service iptables save //保存修改的 iptables 规则
```

配置 SELinux，修改配置文件/etc/selinux/config。

```
SELINUX=permissive  //表示系统会收到警告信息，但是不会受到限制，作为 SELinux 的 debug
                    模式
```

（6）配置 yum 源

将提供的安装光盘和安装文件复制到系统内部，制作安装源，本次测试采用实验室本地源。

① 安装源子目录。

将本书提供的光盘镜像文件 XianDian-IaaS-v1.4.iso 和 CentOS-6.5-x86_64-bin.iso 上传到两个节点服务器自己指定的子目录中（如子目录/var 中）并解压。创建子目录/software，如图 2-6 所示。将子目录/iaas-repo 移至/var/software 子目录中；创建子目录/centos6.5，将子目录/Packages 和/repodata 移至子目录/centos6.5 中，可以删除复制在/var 下的原解压文档。如果是用 VMware 创建的虚拟机来搭建 IaaS 平台，可以通过挂载镜像文档来复制上述软件包。

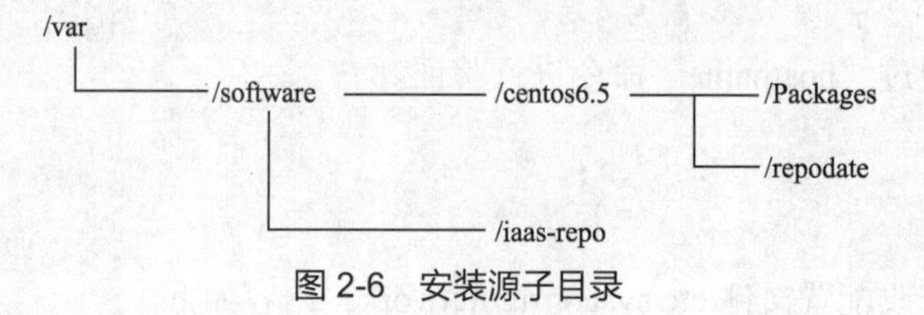

图 2-6　安装源子目录

② 建立 FTP 服务。

进入图 2-6 的子目录，用“rpm”命令安装 FTP 服务器，修改 FTP 默认的文件服务路径为/var/software，即指向存放 yum 源的路径，需在/etc/vsftpd 下修改 vsftpd.conf 配置文件，添加如下配置项。

```
    local_root=/var/software
    chroot_local_user=YES
    anon_root=/var/software
```

存盘后再重启 FTP 服务。

③ yum 源备份。

```
# mv /etc/yum.repos.d/*/opt/
```

④ 创建 repo 文件。

在控制节点的/etc/yum.repos.d/路径下创建 local.repo 文件，内容如下。

```
[centos]
name=centos                                   //设置此 yum 的资源描述名称
baseurl=ftp://172.24.2.10/centos6.5/          //设置 yum 源的访问地址及路径
gpgcheck=0                                    //不使用 gpg 检查 gpgkey
enabled=1                                     //启动此 yum 源
[openstack]
name=openstack
baseurl=ftp://172.24.2.10/iaas-repo/
gpgcheck=0
enabled=1
```

做好上述配置后建议进行 FTP 服务器和“yum”命令的测试，如用“yum”命令安装简单的服务等。

在实例节点的/etc/yum.repos.d/路径下创建 local.repo 文件的过程与上述类似，仅需修改 FTP 服务器的 IP 地址为 172.24.2.20，也可以用控制节点的 FTP 服务器作为 yum 源的资源包。

（7）配置 IP

配置临时 IP，可方便运行安装脚本，以及修改服务器 eth0 端口的地址。

```
# 控制节点
```

修改配置文件/etc/sysconfig/network-scripts/ifcfg-eth0，内容如下。

```
DEVICE=eth0
TYPE=Ethernet
ONBOOT=yes
NM_CONTROLLED=no
BOOTPROTO=static
IPADDR=172.24.2.10
NETMASK=255.255.255.0
GATEWAY=172.24.2.1
```

对配置文件/etc/sysconfig/network-scripts/ifcfg-eth1，只需将上述的内容进行如下修改。

```
IPADDR=172.24.3.10
```

删除 GATEWAY 配置项。

修改完成，重启网络，内容如下。

```
# service network restart
```

```
# 实例节点
```

修改配置文件/etc/sysconfig/network-scripts/ifcfg-eth0，内容如下。

```
DEVICE=eth0
TYPE=Ethernet
ONBOOT=yes
NM_CONTROLLED=no
BOOTPROTO=static
IPADDR=172.24.2.20
NETMASK=255.255.255.0
GATEWAY=172.24.2.1
```

对配置文件/etc/sysconfig/network-scripts/ifcfg-eth1，只需将上述的内容进行如下修改。

```
IPADDR=172.24.3.20
```

删除 GATEWAY 配置项。

修改完成，重启网络。

```
# service network restart
```

（8）重启设备

完成配置后，重启两个节点服务器。

（9）部署脚本安装平台

将提供的安装脚本复制到 CentOS 6.5 系统中。

```
附录一 Xiandian_Pre.sh
附录二 Xiandian_Install_Controller_Node.sh
附录三 Xiandian_Install_Compute_Node.sh
```

（10）配置环境变量

修改附录一 Xiandian_Pre.sh 中各项参数，参数说明如下。

```
Mysql_Admin_Passwd=000000                          //数据库用户密码
Admin_Passwd=000000                                //管理员密码
Demo_User_Passwd=000000                            //演示用户密码
Demo_DB_Passwd=000000                              //演示数据库密码
Contoller_Hostname=controller                      //控制节点主机名
Controller_Mgmt_IPAddress=172.24.2.10              //控制节点管理网段密码
Gateway_Mgmt=172.24.2.1                            //管理网段网关
Controller_External_IPAddress=172.24.3.10          //外部地址
Network_Start_Vlan_ID=43                           //网络节点开始 Vlan ID
Network_End_Vlan_ID=46                             //网络节点结束 Vlan ID
Compute_Hostname=compute                           //实例节点主机名
Compute_Mgmt_IPAddress=172.24.3.20                 //实例节点管理地址
Compute_External_IPAddress=172.24.3.20             //实例节点外部地址
Stroage_Cinder_Disk=sda3                           //Cinder 存储磁盘分区名称
Stroage_Swift_Disk=sda2                            //Swift 存储磁盘分区名称
```

修改完成之后保存配置。

（11）配置控制节点

配置完成环境变量之后，控制节点执行./Xiandian_Install_Controller_Node.sh，在执行过程中单击【Enter】键完成密钥创建，同时分别输入节点密码，完成密钥验证。

（12）配置实例节点

完成环境变量配置之后，实例节点执行./Xiandian_Install_Compute_Node.sh，完成实例节点安装。

2. 验证安装基础工作是否完成

上述操作完成后，打开网页 http://172.24.2.10/dashboard 进行验证服务，若看到图 2-7 所示的 Dashboard 登录界面，并且在使用管理员账号和密码登录后，看到图 2-8 所示的 Dashboard 管理界面，则表示安装基础工作正确完成。

图 2-7　Dashboard 登录界面

图 2-8　Dashboard 管理界面

提示

在操作云平台系统安装基础工作过程中要注意正确使用 vi 文本编辑工具，以及配置文件所在的目录位置，命令操作时要注意命令的英文字母的大小写，Linux 中命令严格区分字母的大小写，修改 Xiandian_Pre.sh 的配置环境变量参数时更应该小心，不能写错。

项目三 认证服务

在 OpenStack 框架中，Keystone（OpenStack Identity Service）的功能是验证身份、校验服务规则和发布服务令牌，它实现了 OpenStack 的 Identity API。Keystone 可分解为两个功能，即权限管理和服务目录。权限管理主要用于用户的管理授权。服务目录类似一个服务总线，或者说是整个 OpenStack 框架的注册表。认证模块提供 API 服务、Token 令牌机制、服务目录、规则和认证发布等功能。

学习目标

- 了解 Keystone 的基本概念。
- 理解 Keystone 的服务流程。
- 掌握租户、用户的不同创建方法。
- 掌握不同的绑定权限的方法。

任务一 Keystone 管理认证用户

任务要求

小李是某公司云计算助理工程师，在公司已经部署好的云计算平台下，小李要学习如何配置 Keystone 认证服务，并学习为公司员工创建用户账号、管理用户权限，具体要求如下。

- 配置并启用认证服务。
- 创建用户账号“alice”。
- 创建租户账号“acme”，用于管理一组账户。
- 创建角色“compute-user”，用于用户权限的管理。
- 绑定用户和租户的权限。

相关知识

1. 相关概念

（1）认证（Authentication）

认证是确认允许一个用户访问的进程。为了确认请求，OpenStack Identity 会为访问用户提供证书，起初这些证书是用户名和密码，或用户名和 API key。当 OpenStack Identity 认证体系接受了用户的请求之后，它会发布一个认证令牌（Token），用户在随后的请求中使用这个令牌去访问资源中其他的应用。

（2）证书（Credentials）

证书是用于确认用户身份的数据，如用户名、密码、API key，或认证服务提供的认证令牌。

（3）令牌（Token）

令牌通常指的是一串比特值或者字符串，用来作为访问资源的记号。Token 中含有可访问资源的范围和有效时间，一个令牌是一个任意比特的文本，用于与其他 OpenStack 服务来共享信息，Keystone 以此来提供一个 central location，以验证访问 OpenStack 服务的用户。令牌的有效期是有限的，可以随时被撤回。

（4）租户（Tenant）

Tenant 即租户，早期版本又称为 Project，它是各个服务中的一些可以访问的资源的集合。例如，通过 Nova 创建虚拟机时要指定到某个租户中，在 Cinder 创建卷时也要指定到某个租户中，用户访问租户的资源前，必须与该租户关联，并且指定该用户在该租户下的角色。

平台构建完毕后会产生“admin”“service”和“demo”三个租户。在这些租户中，“admin”租户代表管理组，拥有平台的最高权限，可以更新、删除和修改系统的任何数据。“service”租户代表平台内所有服务的总集合，平台安装的所有服务默认会被加入到此租户中，为后期的统一管理提供帮助，此租户可以修改当前租户下所有服务的配置信息，提交以及修改租户的内容。“demo”租户则是一个演示测试租户，没有实际的用处。

（5）用户（User）

使用服务的用户，可以是人、服务，或系统使用 OpenStack 相关服务的一个组织。根据不同的安装方式，一个用户可以代表一个客户、账号、组织或项目。用户通过 Keystone Identity 认证登录系统并调用资源。用户可以被分配到特定项目并执行项目相关操作。需要特别指出的是，OpenStack 通过注册相关服务用户来管理服务，例如，Nova 服务注册 Nova 用户来管理相应的服务。对于管理员来说，需要通过 Keystone 来注册管理用户。

（6）角色（Role）

Role 即角色，Role 代表一组用户可以访问的资源权限，如 Nova 中的虚拟机、Glance 中的镜像。Users 可以被添加到任意一个全局的 Role 或租户内的 Role 中。在全局的 Role 中，用户的 Role 权限作用于所有的租户，即可以对所有的租户执行 Role 规定的权限。在租户内的 Role 中，用户仅能在当前租户内执行 Role 规定的权限。

使用云服务的用户不局限于人，也可以是系统或者服务。用户可以通过指定的令牌登录系统并调用资源。用户可以被分配到特定项目并执行项目的相关操作。

平台构建完毕后，系统会创建“_member_”“admin”两个角色，在系统中，“_member_”角色表示系统的普通用户的权限，拥有系统的正常使用和对当前租户的管理权限。“admin”角色是代表系统的管理员身份，对系统有绝对的管理权限。

提示

OpenStack 中项目（Project）、用户（User）和角色（Role）3 者的关系为，项目是用户的集合，项目又称为租户或 accounts，用户可以属于一个或多个项目，角色决定了用户的权限，可以分配角色给 user-project 组。

2. 认证服务流程

用户请求云主机的流程涉及认证 Keystone 服务、计算 Nova 服务、镜像 Glance 服务，在服务流程中，令牌（Token）作为流程认证进行传递，具体服务申请认证机制流程如图 3-1 所示。

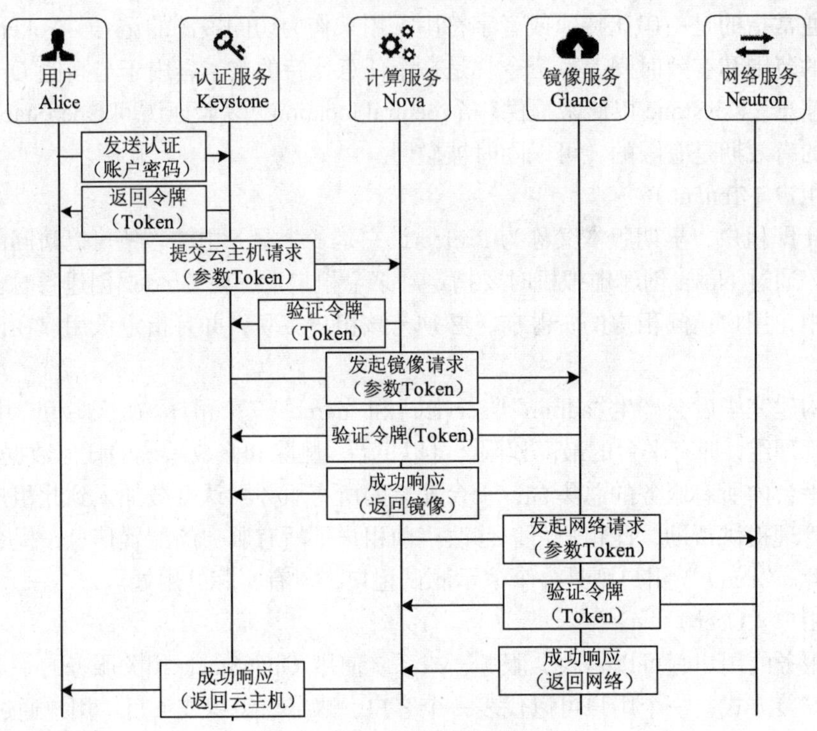

图 3-1　服务申请认证机制流程

任务实现

1. 配置 Keystone 应用环境

在安装 Keystone 服务之前需要指定用户名和密码，通过认证服务来进行身份认证，在开始阶段是没有创建任何的用户的，所以必须使用授权令牌和服务的访问接口来创建特定的用来进行身份认证的用户，之后需要创建一个管理用户的环境变量（admin-openrc.sh）来管理最终的凭证和终端。

在安装Keystone服务之后，产生的主配置文件存放在/etc/keystone目录中，名为keystone.conf，在配置文件中需要配置初始的 Token 值和数据库的连接地址。

Keystone 服务安装完毕后，可以通过请求身份令牌的方式来验证服务，具体命令如下。

```
$keystone --os-username=admin --os-password=000000
--os-auth-url=http://172.24.2.10:35357/v2.0  token-get
//以 admin 用户访问 http://172.24.2.10:35357/v2.0 地址获取 Token 值
```

2. 管理认证用户

OpenStack 的用户(User)包括云平台使用者、服务以及系统。用户可通过认证登录系统并调用资源。为方便管理，用户被分配到一个或多个租户（Tenant）中，租户是用户的集合。为给用户分配不同的权限，Keystone 设置了角色（Role），角色代表用户可以访问的资源等权限。用户可以被添加到任意一个全局的或租户内的角色中。在全局的角色中，用户的角色权限作用于所有的用户，即可以对所有的用户执行角色规定的权限；在租户内的角色中，用户仅能在当前租户内执行角色规定的权限，下面介绍几种常见操作。

（1）创建用户

创建一个名称为"alice"的账户，密码为"mypassword123"，邮箱为"alice@example.com"。命令如下。

```
$keystone user-create --name=alice --pass=mypassword123 --email=alice@example. com
```

具体执行结果如下。

Property	Value
email	alice@example.com
enabled	True
id	76532276ee7a45dfa337a1b5d3667cdb
name	alice
username	alice

从上面的操作可以看出，创建用户需要用户名称、密码和邮件等信息，具体命令格式如下。

```
$ keystone user-create --name <user-name> [--tenant <tenant>]
[--pass [<pass>]] [--email <email>]
[--enabled <true|false>]
```

其中，参数<user-name>代表新建用户名，参数 <tenant>代表绑定租户名。

（2）创建租户

一个租户就是一个项目、团队或组织，当请求 OpenStack 服务时，必须定义一个租户。例如，查询计算服务正在运行的云主机实例列表。

创建一个名为"acme"的租户。

```
$ keystone tenant-create --name=acme
```

执行结果如下。

Property	Value
description	
enabled	True
id	f434deda03814b369be53a346ee00401
name	acme

从上面的操作可以看出，创建租户需要租户名等相关信息，具体命令格式如下。

```
$ keystone tenant-create --name <tenant-name>
                         [--description <tenant-description>]
                         [--enabled <true|false>]
```

其中，参数<tenant-name>代表新建租户名，参数 <tenant-description>代表租户描述名。

（3）创建角色

角色限定了用户的操作权限。例如，创建一个角色“compute-user”。

```
$ keystone role-create --name=compute-user
```

具体执行结果如下。

Property	Value
id	c4c6cfae9848489f841966ac3172afaa
name	compute-user

从上面的操作可以看出，创建角色需要角色名称信息，具体命令格式如下。

```
$ keystone role-create --name <role-name>
```

其中，参数<role-name>代表角色名称。

（4）绑定用户和租户权限

添加的用户需要分配一定的权限，这就需要把用户关联绑定到对应的租户和角色上。例如，给用户“alice”分配“acme”租户下的“compute-user”角色，命令如下。

```
$ keystone user-role-add --user=alice --role=compute-user --tenant-id=
7542b75f42014c1485edbddb2c1cab37
```

执行结果如下。

id	name	enabled	email
db21d0b623e84e48bb1f92b281dcbfa3	admin	True	
76532276ee7a45dfa337a1b5d3667cdb	alice	True	alice@example.com
1b78277a07ff4b9090c531c365eacd69	ceilometer	True	
ae2f12c0b3d84589875c35cb74598cc1	cinder	True	
e7d2d41a39e84456880d4b8a8e9d1f9a	glance	True	
7fdcdc8db7534ae4857fda68ff5b0d44	heat	True	
f4fa21d6f93f4e699b9426b653bfec11	neutron	True	
920457de528a4e9dada1005b9d86a44e	nova	True	
dc109ecfc6b244d4a3baba2a6a73fe2d	sahara	True	
523ae7931fbf4149ba405fc30de8265b	swift	True	

从上面的操作可以看出，绑定用户权限需要用户名称、角色名称和租户名称等信息，具体命令格式如下。

```
$ keystone user-role-add --user <user> --role <role> [--tenant <tenant>]
```

其中，参数<user>代表需要绑定的用户名称，参数<role>代表用户绑定的角色名称，参数<tenant>代表用户绑定的租户名称。

（5）图形化界面操作

以管理员身份登录到 Dashboard 界面，选择“管理员”→“认证面板”→“项目”，可以看到项目（租户）列表，如图 3-2 所示。

图 3-2　云平台管理界面

单击图 3-2 所示界面中的【修改用户】按钮，进入“acme”项目中。在这个界面中，可以查看刚加入的用户“alice”，把该用户分配给用户对应的角色“compute-user”，如图 3-3 所示。完成该操作后，用户“alice”就可用“alice”用户名和对应密码“mypassword123”登录云平台了。

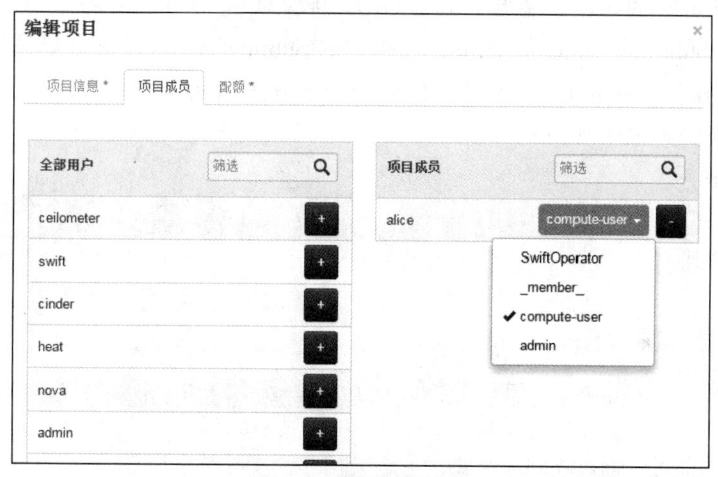

图 3-3　编辑项目窗口

提示

在 OpenStack 中验证服务的身份令牌也可以直接在 admin-openrc.sh 文件中定义系统用户、密码以及认证服务的 Endpoint 等参数，在实际应用中，直接引用（source）环境变量，即可使用 Keystone。

任务二　创建租户、用户并绑定用户权限

任务要求

经过一系列学习之后，小李已经初步掌握到 Keystone 认证服务的使用方法，现在小李要为公司的员工创建相应的部门租户，为员工创建员工用户，并赋予相应的权限。

公司有 100 名员工，其中 50 名为项目研发部（研发环境）员工，45 名为业务部（办公环境）员工，5 名为 IT 工程部（运维环境）员工。

根据企业人员部门分配，现构建 3 个租户、100 个用户，管理人员拥有管理员权限，其余人员拥有普通用户权限，规划如表 3-1 所示。

表 3-1 部门分配规划

部门	租户	用户	权限
项目研发部门	RD_Dept	rduser001 ~ rduser050	普通用户
业务部门	BS_Dept	bsuser001 ~ bsuser045	普通用户
IT 工程部门	IT_Dept	ituser001 ~ ituser005	普通用户和管理员用户

相关知识

OpenStack 服务（Service）包括 Nova、Glance、Swift、Heat、Ceilometer、Cinder 等。Nova 提供云计算服务，Glance 提供镜像管理服务，Swift 提供对象存储服务，Heat 提供资源编排服务，Ceilometer 提供告警计费服务，Cinder 提供块存储服务。为了方便用户调用这些服务，OpenStack 为每一个服务提供一个用于访问的端点（Endpoint），如果需要访问服务，则必须知道它的端点，端点一般为 URL，我们知道服务的 URL，就可以访问它。端点的 URL 具有 public、private 和 admin 三种权限。public URL 可以被全局访问，private URL 只能被局域网访问，admin URL 被从常规的访问中分离出来。

常用的服务管理命令如下。

- 创建服务

```
$ keystone service-create
```

功能：创建服务。

格式：

```
$ keystone service-create --name <name> --type <type>
                          [--description <service-description>]
```

参数说明。

--name <name>：创建的服务名称。

--type <type>：创建的服务类型。

[--description <service-description>]：创建的服务的描述。

- 创建服务访问端点

```
$ keystone endpoint-create
```

功能：创建服务访问的 API 端点。

格式：

```
$ keystone endpoint-create [--region <endpoint-region>] --service
                           <service> --publicurl <public-url>
                           [--adminurl <admin-url>]
                           [--internalurl <internal-url>]
```

参数说明。

[--region <endpoint-region>]：创建端点的区域名称。

--service <service>：创建端点的服务名称。

--publicurl <public-url>：对外服务的 URL 地址。

[--adminurl <admin-url>]：管理网络访问的 URL 地址。

[--internalurl <internal-url>]：内部访问的 URL 地址。

- 查询服务目录

```
$ keystone catalog
```

Service Catalog（服务目录）是 Keystone 为 OpenStack 提供的一个 REST API 端点列表，并以此作为决策参考。

```
$ keystone catalog                                  # 可以显示所有已有的 service
$ keystone catalog --service <service-type>         # 显示某个 service 的信息
```

- 查询 Keystone 服务器和授权协议

```
$ keystone discover
```

发现 Keystone 服务器和授权协议。

```
Keystone found at http://172.24.0.10:35357/v2.0
   - supports version v2.0 (stable) here http://172.24.0.10:35357/v2.0/
      - and s3tokens: OpenStack S3 API
      - and OS-EP-FILTER: OpenStack Keystone Endpoint Filter API
      - and OS-FEDERATION: OpenStack Federation APIs
      - and OS-KSADM: OpenStack Keystone Admin
      - and OS-SIMPLE-CERT: OpenStack Simple Certificate API
      - and OS-EC2: OpenStack EC2 API
```

- 其他常用的 Keystone 的命令

```
$ keystone bash-completion
```

输出可选的命令，即选项。

```
--enabled --tenant_id --value --role help --region tenant-get --user-id user-
list discover ec2-credentials-create --tenant-id --role-name user-role-add
--pass user-delete tenant-delete endpoint-delete --service-id --service_id role-
create endpoint-create password-update --tenant-name service-create--user- name
tenant-update --endpoint-type --new-password -h user-create --tenant --service
--description --wrap endpoint-list ec2-credentials-delete --role_id user-role-
remove role-get tenant-list ec2-credentials-list user-get --user --publicurl
catalog --user_id user-role-list role-delete --endpoint_ type --attr user-update
endpoint-get --type --access ec2-credentials-get --name --internalurl --email
bootstrap role-list user-password-update --help tenant-create --current-
password token-get --adminurl service-delete service-get service-list --role-id
```

任务实现

1. 创建租户

创建项目研发部（Research and Development Department）名为“RD_Dept”的租户、业务部（Business Department）名为“BS_Dept”的租户、IT 工程部（Engineering Department）名为“IT_Dept”的租户。

① 通过 Dashboard 界面为研发部创建一个名为“RD_Dept”的租户。

- 进入 Dashboard 界面找到管理员选项。
- 选择“创建项目”，在弹出窗口中输入“名称”和“描述”信息。在默认情况下，项目是自动激活的，如图 3-4 所示。
- 在“配额”选项卡中，可进行项目资源分配。
- 创建成功后，在项目列表中，会显示出该项目条目，并获得一个自动分配的 ID，如图 3-5 所示。

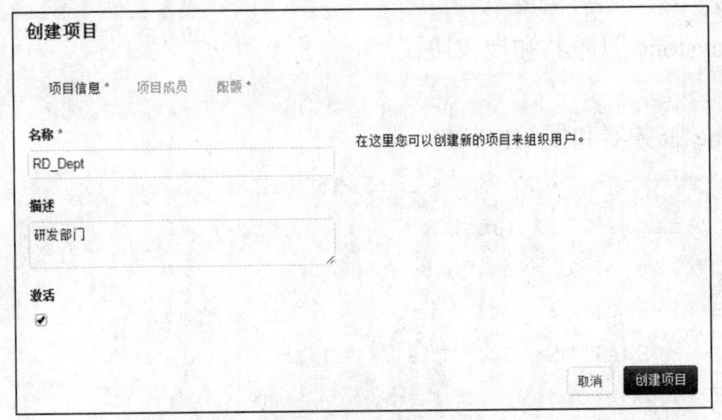

图 3-4　Web 界面创建租户“RD_Dept”

名称	描述	项目ID	激活	动作
acme	-	f434deda03814b369be53a346ee00401	True	修改用户 更多
admin	Admin Tenant	a3c430debe7e48d4a2f58c96dd89746b	True	修改用户 更多
RD_Dept	研发部门	67d4b1d487d84c2a8216a2d337d26a29	True	修改用户 更多
service	Service Tenant	6a280a0b89984078adc4cebc74479f07	True	修改用户 更多

显示4个条目

图 3-5　Web 界面显示租户列表

② 通过 Shell 界面为业务部创建一个名为“BS_Dept”的租户。

```
# keystone tenant-create --name BS_Dept --description 业务部门
                                    //创建业务部门 BS_Dept
```

执行结果如下。

Property	Value
description	业务部门
enabled	True
id	fda7a24a566c40918f25bb963a3decd5
name	BS_Dept

```
# keystone tenant-get BS_Dept                    //获取租户详细信息
```

执行结果如下。

Property	Value
description	业务部门
enabled	True
id	fda7a24a566c40918f25bb963a3decd5
name	BS_Dept

③ 通过脚本为工程部创建一个名为“IT_Dept”的租户（具体脚本参见附录四Keystone-manage-tenant.sh）。

执行 Keystone-manage-tenant.sh 脚本来创建“IT_Dept”租户，在执行脚本前，需要确保该脚本具有执行权限。如果没有执行权限，需要通过“chmod”给文件添加执行权限。

通过“./”或者“/bin/bash”执行脚本文件，输入部门名称和部门描述来创建“IT_Dept”租户。

具体执行命令如下。

```
# ./Keystone-manage-tenant.sh
```

或：

```
# /bin/bash Keystone-manage-user.sh
```

该命令执行结果如下。

```
Please Input new tenant name : eg (openstack)
IT_Dept
Please Input tenant description : eg (openstack description)
IT 工程部门
```

Property	Value
description	业务部门
enabled	True
id	fda7a24a566c40918f25bb963a3decd5
name	BS_Dept

```
Keystone All Tenant List
```

id	name	enabled
ece0ab9d8bed4683b5915cb9f7e595ff	BS_Dept	True
6d22e598c8164e398fb06e87cd08c355	IT_Dept	True
67d4b1d487d84c2a8216a2d337d26a29	RD_Dept	True
f434deda03814b369be53a346ee00401	acme	True
a3c430debe7e48d4a2f58c96dd89746b	admin	True
6a280a0b89984078adc4cebc74479f07	service	True

2. 创建用户账号

为项目研发部创建50个用户，分别名为“rduser001”~“rduser050”，密码为“cloudpasswd”，为业务部创建45个用户，分别名为“bsuser001”~“bsuser045”，密码为“cloudpasswd”，

为 IT 工程部创建 5 个用户，分别名为“ituser001” ~ “ituser005”，密码为“cloudpasswd”。可使用 GUI 和 CLI 界面，辅助使用 Shell。

① 通过 Dashboard 界面为项目研发部创建用户“rduser001”，密码为“cloudpasswd”。

- 进入 Dashboard 界面找到管理员选项。
- 打开认证面板，选中“用户”。
- 选择“创建用户”。
- 在“创建用户”界面，输入“用户名”“邮箱”“密码”“主项目”和“角色”，如图 3-6 所示。

图 3-6　Web 界面创建用户“rduser001”

结果如图 3-7 所示。

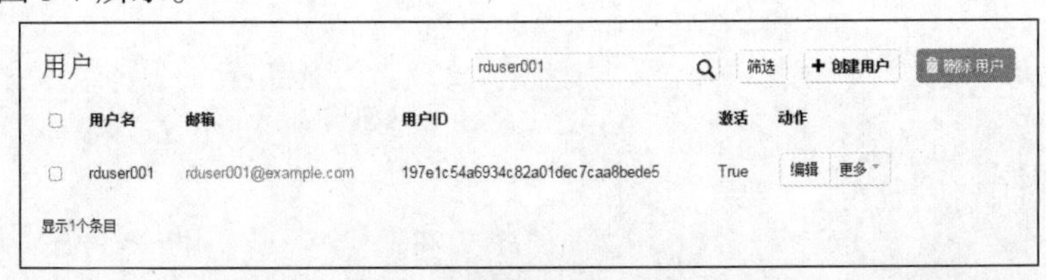

图 3-7　Web 界面创建用户“rduser001”

② 通过 Shell 命令行为项目研发部创建用户“rduser002”，密码为“cloudpasswd”。

```
# keystone user-create --name rduser002 --pass cloudpasswd --email rduser002
@example.com
```

执行结果如下。

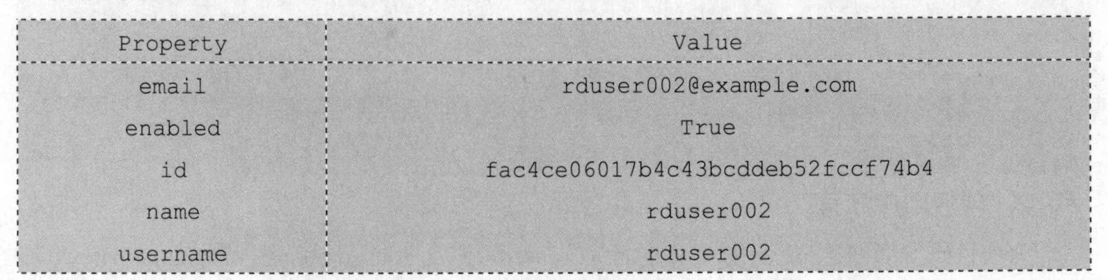

Property	Value
email	rduser002@example.com
enabled	True
id	fac4ce06017b4c43bcddeb52fccf74b4
name	rduser002
username	rduser002

③ 通过执行 Shell 脚本 Keystone-manage-user.sh 为项目研发部创建用户“rduser003”

～“rduser050”，密码为“cloudpasswd”。（具体脚本参见附录五 Keystone-manage-user.sh）

在命令行内输入# ./Keystone-manage-user.sh，执行该脚本。在命令行内按提示输入用户名称、用户密码、电子邮件域名地址、用户角色（这里只能赋予一个角色）和用户所属部门。

执行命令后结果如下所示。

```
Please Input New User Name : eg (username)
rduser
Please Input User Password: eg (000000)
000000
Please Input User Email Address,If don't need  press enter: eg (openstack.com)
example.com
Please Input User  Beginning And End  Number: eg (001-002)
003-050
Please enter the User belong Roles Name, Press enter for '_member_' role by default:
eg (admin)

Please Input User belong Tenant Name: eg (tenantname)
RD_Dept
```

Property	Value
email	rduser003@example.com
enabled	True
id	9d81d5e551124ccdb8f0967ad6d0d047
name	rduser003
username	rduser003

同理，可通过执行 Shell 脚本为业务部创建用户“bsuser001”～“bsuser045”，密码为“cloudpasswd”。执行命令后结果如下所示。

```
Please Input New User Name : eg (username)
bsuser
Please Input User Password: eg (000000)
000000
Please Input User Email Address,If don't need  press enter: eg (openstack.com)
example.com
Please Input User  Beginning And End  Number: eg (001-002)
001-045
Please enter the User belong Roles Name, Press enter for '_member_' role by default:
eg (admin)

Please Input User belong Tenant Name: eg (tenantname)
BS_Dept
```

Property	Value
email	bsuser001@example.com
enabled	True
id	dafde51965ea4e8f99bc0ce4ee4eff67
name	bsuser001
username	bsuser001

Property	Value
email	bsuser002@example.com
enabled	True
id	43c9f76d72f7445d83e73cc7ba76b194
name	bsuser002
username	bsuser002

④ 通过执行 Shell 脚本为 IT 工程部创建用户"ituser001"~"ituser005"，密码为"cloudpasswd"。执行过程和结果如下所示。

```
Please Input New User Name : eg (username)
ituser
Please Input User Password: eg (000000)
000000
Please Input User Email Address,If don't need  press enter: eg (openstack.com)
example.com
Please Input User   Beginning And End  Number: eg (001-002)
001-005
Please enter the User belong Roles Name, Press enter for '_member_' role by default:
eg (admin)

Please Input User belong Tenant Name: eg (tenantname)
IT_Dept
```

Property	Value
email	ituser002@example.com
enabled	True
id	c0b0fc106caa469f87103992344a790c
name	ituser002
username	ituser002

3. 绑定用户权限

将项目研发部、业务部的用户绑定普通用户权限；将 IT 工程部的用户绑定管理员和普通用户权限。

① 通过 Dashboard 界面将项目研发部用户"rduser001"绑定普通用户权限。

- 进入 Dashboard 界面找到管理员选项。

- 打开认证面板，选中“项目”。
- 找到相应的项目，在动作栏目中选择“修改用户”，进入到“编辑项目”对话框。
- 在“项目成员”中，为项目用户选择相应的角色，如图 3-8 所示。

图 3-8 Web 界面修改用户权限

② 通过 Shell 命令行将项目研发部用户“rduser002”绑定普通用户权限。

命令如下。

```
# keystone user-role-add --user rduser002 --tenant RD_Dept --role _member_
# keystone user-role-list --user rduser002 --tenant RD_Dept
```

部分执行结果如下所示。

id	name	user_id	tenant_id
9fe2ff9ee4384b1894a90878d3e92bab	_member_	fac4ce06017b4c43bcddeb52fccf74b4	67d4b1d487

③ 通过执行 Keystone-manage-add-role.sh 脚本将项目 IT 工程部用“ituser001”～“ituser005”绑定普通用户和管理员用户权限，部分执行结果如下所示。（具体脚本参见附录六 Keystone-manage-add-role.sh）

```
Please Enter The User Name
ituser
Please Input User  Beginning And End  Number: eg (001-002)
001-001
Please Enter the Tenant Name
IT_Dept
Please Enter the  Role Name
admin
Keystone user ituser001 tenant IT_Dept role list
```

id	name	user_id	tenant_id
9fe2ff9ee4384b1894a90878d3e92bab	_member_	c0b0fc106caa469f87103992344a790c	6d22e598c
dca7d06e0abf43a6a6947fe3e6c84a2e	admin	c0b0fc106caa469f87103992344a790c	6d22e598c

提示

在 OpenStack 用户管理中，用户、角色和租户等身份的权限和等级设定非常重要，对应的身份只拥有对应的适合身份的权限，对系统的安全和稳定非常有帮助。

项目四 基础控制服务

通过认证服务的学习，我们可以以不同的身份访问企业云平台，可以通过项目研发部的账户登录项目研发部，可以通过业务部的帐户访问业务部的资源，也可以通过IT工程部的身份登录查看整个系统的运行状况。下面我们继续学习消息服务（QPID）、镜像服务（Glance）和计算控制服务（Nova），了解这3个组件是如何为平台的正常运行提供支撑的。

学习目标

- 了解QPID、Glance和Nova的基本概念。
- 理解3种服务的服务流程和工作机制。
- 掌握3种服务的基本操作及常见运维。

任务一 消息队列服务

任务要求

在日常的工作生活中，消息传递是一个必不可少的需求。在大型软件的内部信息交换和外部信息传递中，消息传递都是不可或缺的。在系统间通信传递的最基本方法是Socket，但是这是一个最底层的协议，所以在使用时需要程序来调用。

在进行后序的学习过程之前，小李首先要了解消息服务的基本状况和使用的情景，以及OpenStack的RPC（远程呼叫机制）的运行机制。

相关知识

1. 消息队列

AMQP是一种标准化的消息中间件协议，全称为高级消息队列协议（Advanced Message Queuing Protocol），它可以让不同语言、不同系统的应用互相通信，并提供一个简单统一的模型和编程接口。这样，就可以采用各种语言和平台来实现自身的应用，当需要和其他系统通信时，只要承认AMQP即可。

2. QPID消息服务

QPID是Apache开发的一款面向对象的消息中间件，它是AMQP中的一种，可以和其他符合AMQP的系统进行通信。QPID提供了C++/Python/Java/C#等主流编程语言的客户端库，安装使用非常方便。QPID社区十分活跃，有望成为标准AMQP中间件产品。除了符合AMQP基本要求之外，QPID提供了很多额外的（如HA）特性，非常适合集群环境下的消息通信。

提示

了解了 OpenStack 中的消息服务机制才能更好地深入学习后面知识。

任务实现

1. 了解消息队列 AMQP

图 4-1 所示为消息队列 AMQP 服务架构图。

AMQP 中有 3 个重要的角色，如图 4-2 所示。

- Publisher：消息的发送者。
- Exchange：消息的传递者。
- Consumer：消息的接收者。

以模拟写信来说明其工作的方式。为了传递给收件人，首先需要用信封把信的内容装起来，然后在信封上写好收件人的信息，再把信放到邮筒里，后面邮局会拿到信然后根据信封上的收件人信息来看最终把信给谁。写信的人就是 Publisher，邮局就是 Exchange，收件人就是 Consumer。

图 4-1 AMQP 架构图

不同的 Consumer 会创建不同的 Queue（消息队列），然后将对应的 Exchange 绑定到 Queue 上。在消息的传递过程中，Publisher 不会直接地把 Message 放到 Queue 中，也不管 Message 如何分发，Publisher 只管准备好消息，然后交给 Exchange。而 Exchange 做的事情也很简单，一手从 Publisher 拿到消息，然后就把消息放入 Queue 中。

对于 Exchange 来说，是把消息放到某一个 Queue 中，还是放到多个 Queue 中？实际上，在 Publisher 发出消息的时候会附加一个条件，Exchange 会根据这个条件来决定发送的方法，这个条件就是 routingkey。

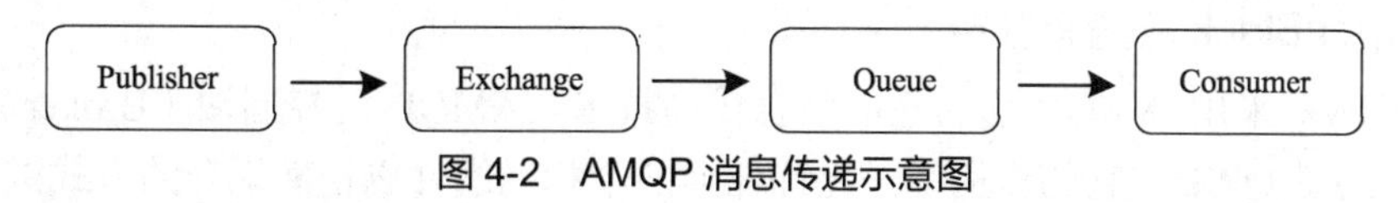

图 4-2 AMQP 消息传递示意图

2. 了解 QPID 消息服务

图 4-3 所示为 QPID 架构图。

在一个采用了消息中间件的通信体系中，有 3 个基本的角色，一个是发送消息的进程，一个是接收消息的进程，他们彼此之间通过 Broker 连接起来，传递消息。

通过基本架构图，用户可知道简单的消息之间的传递方法，接下来用户可以通过 qpid-tool 工具来实际查看其中的实现流程。

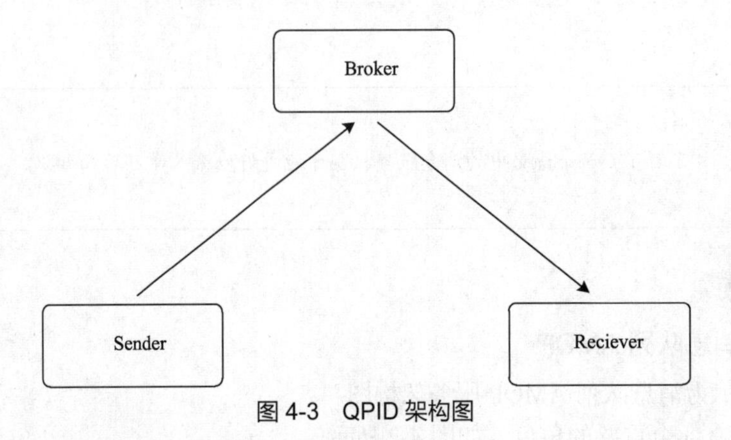

图 4-3 QPID 架构图

QPID 有如下几种命令行工具。

① qpid-config：显示配置、队列和绑定等信息。

② qpid-tool：配置工具，通过配置可以控制 Broker 端。

③ qpid-queue-stats：显示队列信息和队列中的数据情况。

④ qpid-cluster：配置和查看集群。

用户可以通过 qpid-tool 工具进入消息服务中，查看存在的消息情况，如附录一所示。用户可以发现其中与服务相关的 Exchange 有 OpenStack、具体组件_fanout、组件的 Scheduler_fanout、reply_ID。

以 Cinder Volume 为例，Cinder Volume_fanout 负责将 Fanout Publisher 的请求发送到 Cinder Volume_fanout_uuid 消息队列中。

Cinder Scheduler_fanout 负责将 Fanout Publisher 的请求发送到 Cinder Scheduler_fanout_uuid 消息队列中。

OpenStack 负责 Topic Publisher 的请求，根据具体的 routingkey 发送到对应的消息队列中。

Reply_XXX 是由 Direct Publisher 创建的，用于消息处理返回结果通信。

命令如下。

```
# qpid-tool 172.24.2.10
```

输出结果详情见附录七 qpid-tool.txt。

3. OpenStack 的消息服务

OpenStack 采用 AMQP 作为它的消息传递技术。AMQP 的控制端（Broker），不管是 RabbitMQ 还是 QPID，它的作用都是允许任何两个相同组件以松散耦合的方式进行交流。更准确地说，OpenStack 的组件均可以使用远程呼叫机制（RPC）进行彼此沟通。然而，这种范式是建立在"publish/subscribe 模式"（订阅/发布模式，订阅/发布模式定义了一种一对多的依赖关系，可让多个订阅者对象同时监听某一个主题对象。这个主题对象在自身状态变化时，会通知所有订阅者对象，使它们能够自动更新自己的状态）上的，如图 4-4 所示。这样做的好处如下。

① 客户端和服务端之间的解耦（如客户端不需要知道服务引用）。

② 客户端和服务端之间不需要同时使用消息调用，只需要其中一端发送消息调用指令。

③ 随机平衡的远程调用会随机将运行的指令发送到一个节点上。

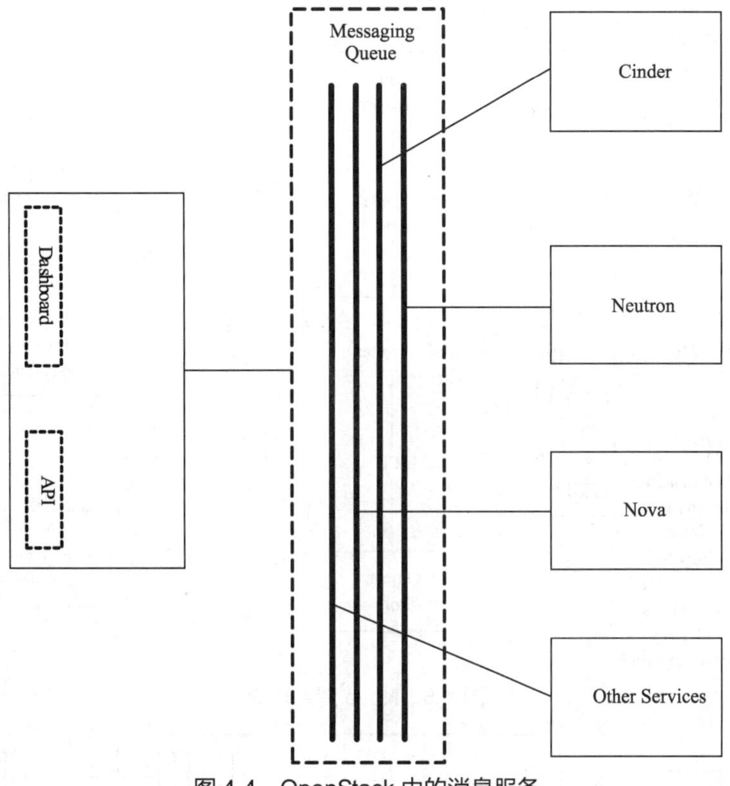

图 4-4　OpenStack 中的消息服务

下面以计算服务 Nova 为例，讲解 OpenStack 组件使用消息服务的原理和机制。Nova 中的每个组件都会连接消息服务器，一个组件可能是一个消息发送者（如 API、Scheduler），也可能是一个消息接收者（如 Compute、Volume、Network）。

发送消息有两种方式：同步调用 rpc.call 和异步调用 rpc.cast。Nova 实现了 RPC（包括请求+响应的 rpc.call 和只有请求的 rpc.cast），Nova 通过 AMQP 提供一个代理类，这个类提供了函数调用时的消息的数据编码和数据解码功能。每个 Nova 服务（如计算）在初始化时创建两个队列，一个接收消息路由的关键节点类型，产生对应的"节点编号"（如 compute.hostname），另一个接收消息路由键返回的节点，生成通用的"节点类型"（如计算）。前者是专门用于 nova api 把这个队列的消息定向到一个特定的节点，在这种情况下，只有主机的计算节点的虚拟机监控程序才能杀死正在运行的虚拟机实例。当使用 rpc.call 时，API 充当 Consumer，当使用 rpc.cast 时，API 只是充当 Publisher，如图 4-5 所示。

4. Nova RPC 映射

当一个实例在 OpenStack 进行云部署和共享时，每个组件通过 Nova 连接到 MessageBroker，根据其个性（如计算节点或网络节点），使用队列作为调用者（如 API 或调度器）或运行组件（如计算或网络）。调用者和组件不实际存在于 Nova 对象模型中，但是在这个例子中使用它们作为一个抽象对象。一个调用程序是一个组件，它使用 RPC 发送消息队列系统，接收消息队列系统，并相应地回复 RPC 调用操作，如图 4-6 所示。

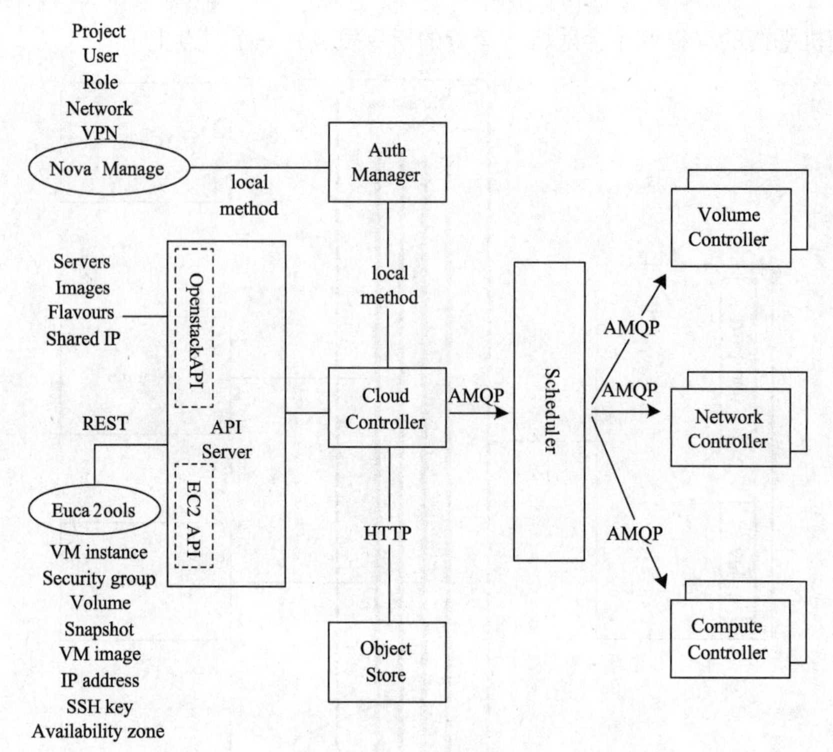

图 4-5 Nova 消息传递

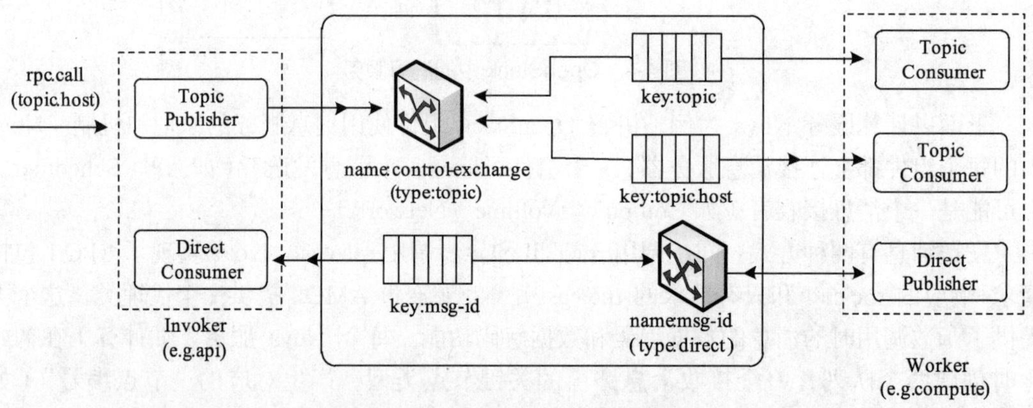

图 4-6 OpenStack RPC 调用机制

下面介绍几个 OpenStack 在消息传递过程中常用的对象。

① Topic Publisher：该对象在进行 rpc.call 或 rpc.cast 调用时创建，每个对象都会连接同一个 topic 类型的交换器，消息发送完毕后对象被回收。

② Direct Publisher：该对象在进行 rpc.call 调用时创建，用于向消息发送者返回响应。该对象会根据接收到的消息属性连接一个 direct 类型的交换器。

③ Direct Consumer：该对象在进行 rpc.call 调用时创建，用于接收响应消息。每一个对象都会通过一个队列连接一个 direct 类型的交换器（队列和交换器以 UUID 命名）。

④ Topic Consumer：该对象在内部服务初始化时创建，在服务过程中一直存在。用于从队列中接收消息，调用消息属性中指定的函数。该对象通过一个共享队列或一个私有队

列连接一个 topic 类型的交换器。每一个内部服务都有两个 Topic Consumer，一个用于 rpc.cast 调用（此时连接的是 bindingkey 为“topic”的共享队列），另一个用于 rpc.call 调用（此时连接的是 bindingkey 为“topic.host”的私有队列）

⑤ Topic Exchange：topic 类型交换器，每一个消息代理节点只有一个 topic 类型的交换器。

⑥ Direct Exchange：direct 类型的交换器，存在于 rpc.call 调用过程中，对于每一个 rpc.call 调用，都会产生该对象的一个实例。

⑦ Queue Element：消息队列。可以共享也可以私有。routingkey 为“topic”的队列会在相同类型的服务中共享（如多个 compute 节点共享一个 routingkey 为“topic”的队列）。

任务二　学习镜像服务

任务要求

小李经过一系列学习之后，已经初步掌握了 OpenStack 消息服务的机制。现在，小李要了解 OpenStack 的另外一种服务——Glance。完成本任务学习后，小李能够了解 Glance 镜像服务在 OpenStack 整体架构的作用，清晰服务框架流程，并熟练掌握镜像的制作方法。

相关知识

1. 概述

Glance 镜像服务可实现发现、注册、获取虚拟机镜像和镜像元数据的功能，镜像数据支持存储多种存储系统，可以是简单文件系统、对象存储系统等。

2. Glance 服务架构

Glance 镜像服务是典型的 C/S 架构，Glance 架构包括 Glance Client、Glance 和 Glance Store。Glance 主要包括 REST API、数据库抽象层（DAL）、域控制器（Glance Domain Controller）和注册层（Registry Layer），Glance 使用集中数据库（Glance DB）在 Glance 各组件间直接共享数据。

所有的镜像文件操作都通过 Glance Store 库完成，Glance Store 库提供了通用接口，对接后端外部的不同存储，如图 4-7 所示。

① 客户端（Client）：外部用于同 Glance 服务的交互和操作。

② Glance API：Glance 对外的 REST 接口。

③ 数据库抽象层（DAL）：Glance 和数据库直接交互的编程接口。

④ Glance 域控制器：中间件实现 Glance 的认证、通知、策略和数据链接等主要功能。

⑤ 注册层：可选层，用于管理域控制和数据库 DAL 层之间的安全通信。

⑥ Glance DB：存储镜像的元数据，根据需要可以选择不同类型的数据库，目前采用 MySQL。

⑦ Glance Store：Glance 对接不同数据存储的抽象层。

⑧ 后端存储：实际接入的存储系统。可以接入简单文件系统、Swift、Ceph 和 S3 云存储等。当前框架选择存储在本地，目录在控制节点/var/lib/glance/images/。

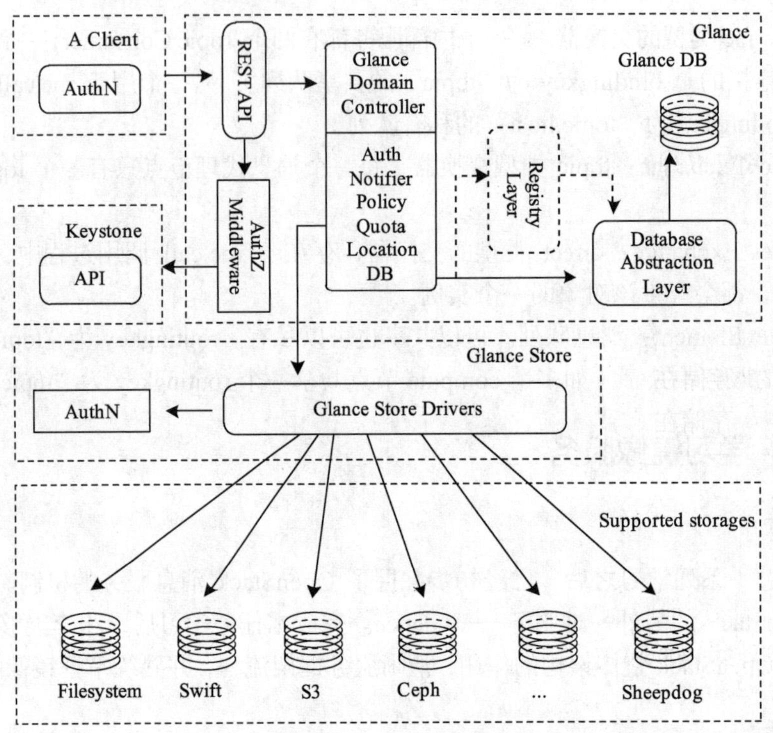

图 4-7　Glance 架构图

Glance 服务自带两个配置文件，在使用 Glance 镜像服务时需要配置 glance-api.conf 和 glance-registry.conf 两个服务模块。在添加镜像到 Glance 时，需要指定虚拟镜像的磁盘格式（Disk Format）和容器格式（Container Format）。

镜像服务运行两个后台服务进程 Demon，如下。

① Glance API：对外接收并发布镜像、获取和存储的请求调用。

② Glance Registry：存储、处理和获取镜像元数据，内部服务只供 Glance 内部使用，不暴露给用户。

③ Glance All：是对前两个进程的通用封装，操作方式和结果一致。

3. 镜像文件格式

虚拟机镜像需要指定磁盘格式和容器格式。虚拟设备供应商将不同的格式布局的信息存在一个虚拟机磁盘映像中，虚拟机的磁盘镜像的基本格式有如下几种。

① RAW：非结构化磁盘镜像格式。

② QCOW2：QEMU 模拟器支持的可动态扩展、写时复制的磁盘格式，是 KVM 虚拟机默认使用的磁盘文件格式。

③ AMI/AKI/ARI：Amazon EC2 最初支持的镜像格式。

④ UEC Tarball：Ubuntu Enterprise Cloud Tarball 是一个经 gzip 压缩后的 tar 文件，包含 AMI、AKI 和 ARI 三种类型的文件。

⑤ VHD ：Microsoft Virtual Hard Disk Format（微软虚拟磁盘文件）的简称。

⑥ VDI：VirtualBox 使用 VDI（Virtual Disk Image）的镜像格式，OpenStack 没有提供直接的支持，需要进行格式转换。

⑦ VMDK：VMWare Virtual Machine Disk Format 是虚拟机 VMware 创建的虚拟机格式。

⑧ OVF：Open Virtualization Format，开放虚拟化格式，OVF 文件是一种开源的文件规范，可用于虚拟机文件的打包。

容器格式可以理解成是虚拟机镜像添加元数据后重新打包的格式，目前有以下几种容器格式。

① BARE：指定没有容器和元数据封装在镜像中，如果 Glance 和 OpenStack 的其他服务没有使用容器格式的字符串，为了安全，建议设置 BARE。

② OVF：OVF 的容器模式。

③ AKI：存储在 Glance 中的是 Amazon 的内核镜像。

④ ARI：存储在 Glance 中的是 Amazon 的 ramdisk 镜像。

⑤ AMI：存储在 Glance 中的是 Amazon 的 machine 镜像。

⑥ OVA：存储在 Glance 中的是 OVA 的 tar 归档文件。

4. 镜像状态

由于镜像文件都比较大，镜像从创建到成功上传到 Glance 文件系统中的过程，是通过异步任务的方式一步步完成的，状态包括 Queued（排队）、Saving（保存中）、Active（有效）、Deactivated（无效）、Killed（错误）、Deleted（被删除）和 Pending_delete（等待删除），如图 4-8 所示。

① Queued 排队：镜像 ID 已经创建和注册，但是镜像数据还没有上传。

② Saving 保存中：镜像数据在上传中。

③ Active 有效：镜像成功创建，状态有效可用。

④ Deactivated 无效：镜像成功创建，镜像对非管理员用户不可用。

⑤ Killed 错误：上传镜像数据出错，目前不可读取。

⑥ Deleted 被删除：镜像不可用，将被自动删除。

⑦ Pending_delete 等待删除：镜像不可用，等待将被自动删除。

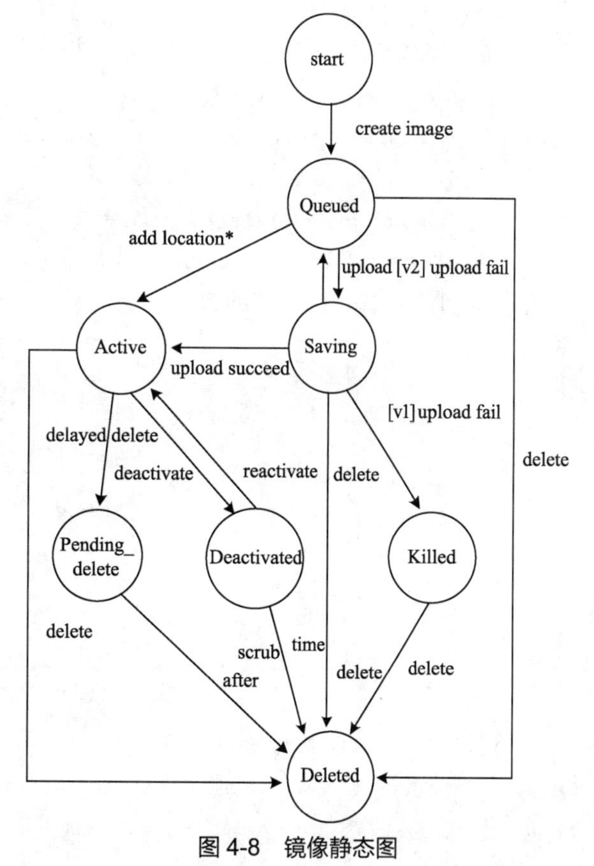

图 4-8　镜像静态图

任务实现

1. 镜像服务基本操作

CirrOS 是一个极小的云操作系统，下面的基础操作将使用这个小系统来进行演示。

查询 Glance 服务状态的命令如下。

```
# glance-control all status
```

查询结果如下所示。

```
glance-api (pid 2633) is running…
glance-registry (pid 2654) is running…
glance_scrubber is stoped
```

查询 Glance API 版本的命令如下。

```
# glance-api --version
```

查询结果如下所示。

```
2014.1.3
```

查询 Glance Control 版本的命令如下。

```
# glance-control --version
```

查询结果如下所示。

```
2014.1.3
```

下面的命令的作用是启动相关服务，并设置为开机启动。

```
# service openstack-glance-api start
# service openstack-glance-registry start
# chkconfig openstack-glance-api on
# chkconfig openstack-glance-registry on
```

在 Dashboard 界面中以“admin”账户登录，可查看镜像。下面将通过从本地上传镜像文件和从网站下载镜像文件的方式创建、查询、删除和修改镜像，如图 4-9 所示。

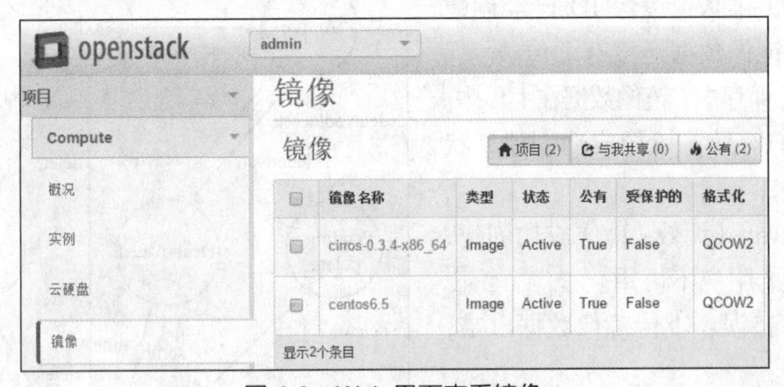

图 4-9 Web 界面查看镜像

（1）下载 CirrOS 镜像文件

```
# mkdir /tmp/images
# cd /tmp/images/
# wget http://download.cirros-cloud.net/0.3.2/cirros-0.3.2-x86_64-disk.img
```

执行结果如下所示。

```
--2016-10-1201:14:05--  http://download.cirros-cloud.net/0.3.2/cirros-0. 3.2-
x86_64-disk.img
Resolving download.cirros-cloud.net... 64.90.42.85, 2607:f298:6:a036:: bd6:a72a
```

```
Connecting to download.cirros-cloud.net|64.90.42.85|:80... connected.
HTTP request sent, awaiting response... 200 OK
Length: 13167616 (13M) [text/plain]
Saving to: "cirros-0.3.2-x86_64-disk.img"
100%[=========================================>] 13,167,616  104K/s  in 5m 42s
2016-10-12 01:19:47 (37.6 KB/s) - "cirros-0.3.2-x86_64-disk.img" saved
[13167616/13167616]
```

下载完成后，查看文件信息，命令如下。

```
# file cirros-0.3.2-x86_64-disk.img
```

执行结果如下所示。

```
cirros-0.3.2-x86_64-disk.img: Qemu Image, Format: Qcow , Version: 2
```

（2）使用命令行创建镜像

```
#glance image-create --name "cirros-0.3.2-x86_64" --disk-format qcow2
--container-format bare \
--is-public True --progress < cirros-0.3.2-x86_64-disk.img
```

执行结果如下所示。

```
[root@controller opt]# glance image-create --name "cirros-0.3.2-x86_64" \
--disk-format qcow2 --container-format bare is-public True --progress <
cirros-0.3.2-x86_64-disk.img
[=============================>] 100%
```

Property	Value
checksum	64d7c1cd2b6f60c92c14662941cb7913
container_format	bare
created_at	2016-10-11T17:28:15
deleted	False
deleted_at	None
disk_format	qcow2
id	9ec221da-9614-43d0-b785-57652c526227
is_public	True
min_disk	0
min_ram	0
name	cirros-0.3.2-x86_64
owner	7daccf6241f2417e81bf6479a310e617
protected	False
size	13167616
status	active
updated_at	2016-10-11T17:28:15
virtual_size	None

创建成功后，可以登录界面，查看镜像信息，如图 4-10 所示。

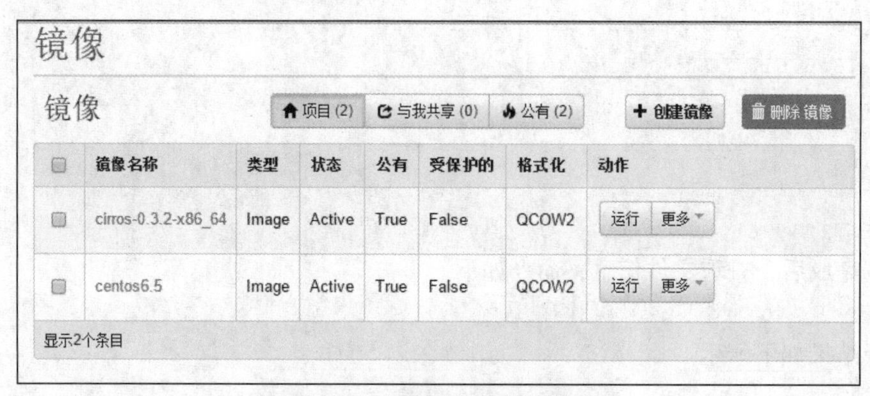

镜像

镜像　　项目 (2)　与我共享 (0)　公有 (2)　　+ 创建镜像　删除镜像

	镜像名称	类型	状态	公有	受保护的	格式化	动作
	cirros-0.3.2-x86_64	Image	Active	True	False	QCOW2	运行 更多
	centos6.5	Image	Active	True	False	QCOW2	运行 更多

显示2个条目

图 4-10　使用界面查看镜像信息

（3）通过 Dashboard 界面创建镜像

通过界面，选择从互联网直接下载，创建镜像。

镜像采用 0.3.4 版本。镜像名称为“cirros-0.3.4-x86_64”，格式选择“QCOW2-QEMU 模拟器”，如图 4-11 所示。

创建一个镜像

名称 *

cirros-0.3.4-x86_64

描述

cirros-0.3.4-x86_64

镜像源

镜像地址

镜像地址

http://download.cirros-cloud.net/0.3.4/cirros-0.3.4-x8

格式化 *

QCOW2 - QEMU 模拟器

构架

最小磁盘（GB）

最小内存(MB)

公有

受保护的

说明：

指定镜像上传到镜像服务

目前只支持HTTP URL可用镜像。镜像服务必须能够访问到镜像地址。支持镜像的二进制压缩格式(.zip,.tar,.gz.)

请注意： 镜像地址必须是有效的直接定位到镜像二进制文件的URL。URL被重定向或者服务器返回错误页面将导致镜像不可用。

取消　创建镜像

图 4-11　通过界面创建镜像

单击“创建镜像”按钮，可以看到新镜像“cirros-0.3.4-x86_64”被创建，状态从“Queued”开始，成功后变为“Active”，创建结果如图 4-12 所示。

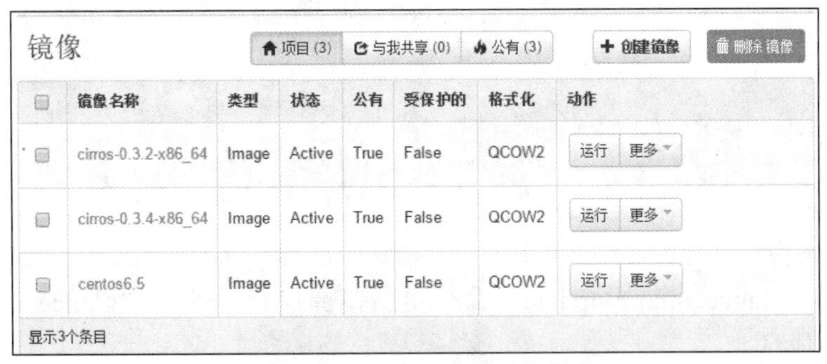

图 4-12　通过界面查看镜像信息

（4）查看镜像信息

下面通过命令方式，查看镜像的基本信息。

先使用“glance image-list”命令，查看镜像列表。

```
# glances image-list
```

查询结果如下所示。

ID	Name	Disk Format	Container Format	Size	Status
f7cac855-e4a1-4690-bd1b-f8d79f90286f	centos6.5	qcow2	bare	305397760	active
9ec221da-9614-43d0-b785-57652c526227	cirros-0.3.2-x86_64	qcow2	bare	13167616	active
5d93b088-0fae-4f3e-af83-5eb4195226f7	cirros-0.3.4-x86_64	qcow2	bare	13287936	active

然后通过“glance image-show”命令查看镜像的详细信息，其中，参数可以是镜像 ID 或者镜像名称，如下所示。

```
# glance image-show f7cac855-e4a1-4690-bd1b-f8d79f90286f
```

查询结果如下所示。

Property	Value
checksum	dfbd01ddbb81c9e8254de236e5e83b0f
container_format	bare
created_at	2016-10-11T06:20:49
deleted	False
disk_format	qcow2
id	f7cac855-e4a1-4690-bd1b-f8d79f90286f
is_public	True
min_disk	0

min_ram	0
name	centos6.5
owner	7daccf6241f2417e81bf6479a310e617
protected	False
size	305397760
status	active
updated_at	2016-10-11T06:20:51

（5）更改镜像

可以使用“glance image-update”命令更新镜像信息，使用“glance image-delete”命令删除镜像信息。

如果需要改变镜像启动硬盘的最低要求值（min-disk）（默认单位为 GB），可使用“glance image-update”命令更新镜像信息，操作如下。

```
# glance image-update --min-disk=1 cirros-0.3.2-x86_64
```

查询结果如下所示。

Property	Value
checksum	64d7c1cd2b6f60c92c14662941cb7913
container_format	bare
created_at	2016-10-11T17:28:15
deleted	False
deleted_at	None
disk_format	qcow2
id	9ec221da-9614-43d0-b785-57652c526227
is_public	True
min_disk	0
min_ram	0
name	cirros-0.3.2-x86_64
owner	7daccf6241f2417e81bf6479a310e617
protected	False
size	13167616
status	active
updated_at	2016-10-11T18:41:11
virtual_size	None

通过命令删除镜像 cirros-0.3.2-x86_64 的操作如下，执行结果如下所示。

```
# glance image-delete cirros-0.3.2-x86_64
# glance image-list
```

ID	Name	Disk Format	Container Format	Size	Status
f7cac855-e4a1-4690-	centos6.5	qcow2	bare	305397760	active

bd1b-f8d79f90286f					
5d93b088-0fae-4f3e-af83-5eb4195226f7	cirros-0.3.4-x86_64	qcow2	bare	13287936	active

2. 制作 Windows 7 镜像

制作一个名为 Cloud_Win7_64bit 的镜像，C 盘大小为 50GB，预装 Office 和 Eclipse 软件，配置 Java 环境（配置名为“CloudUser”的用户，密码为“cloudpasswd”）。

首先要确保 cn_windows_7_ultimate_x64_dvd_x15-66043.iso、virtio-win-0.1-52.iso 和 virtio-win-1.1.16.vfd 文件在服务器中，这里将镜像文件放在/opt 目录中。

（1）安装虚拟化工具软件包

```
# yum install virt-manager libvirt qemu-img virt-viewer
```

（2）修改/etc/libvirt/qemu.conf

```
# vi /etc/libvirt/qemu.conf
```

增加 vnc_listen = "0.0.0.0"，如下所示。

```
[root@controller ~]# vi /etc/libvirt/qemu.conf
# Master configuration file for the QEMU driver.
# All settings described here are optional - if omitted, sensible
# defaults are used.

# VNC is configured to listen on 127.0.0.1 by default.
# To make it listen on all public interfaces, uncomment
# this next option.
#
# NB, strong recommendation to enable TLS + x509 certificate
# verification when allowing public access
#
vnc_listen = "0.0.0.0"
```

（3）重启 Libvirtd

命令如下，执行结果如下所示。

```
# service libvirtd restart
Stopping libvirtd daemon: [ FAILED]
Starting libvirtd daemon: [  OK  ]
```

（4）制作镜像文件

```
# cd /tmp
# qemu-img create -f raw Cloud_Win7_64bit.img 50G
#  virt-install  --name  Cloud_Win7_64bit  --ram  2048  --vcpus  2  --network
network=default, model=virtio --disk=Cloud_Win7_64bit.img, format=raw, device=
disk , bus=virtio  --cdrom=/opt/cn_windows_7_ultimate_x64_dvd_x15- 66043.iso
--disk=/opt/virtio-win-0.1-52.iso --disk=/opt/virtio-win-1.1.16. vfd, device=
```

```
floppy --graphics listen=0.0.0.0, port=5901 --video qxl --channel spicevmc -os
-type windows --os-variant win7 --force
```

（5）打开 VNC

图 4-13 所示为 VNC 登录界面。

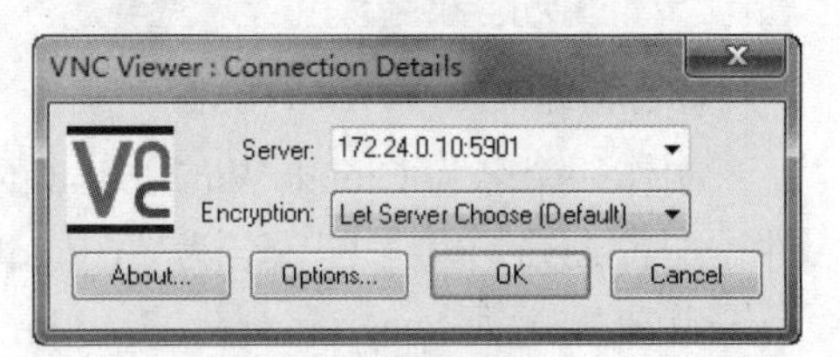

图 4-13　VNC 登录界面

（6）选择要安装的驱动程序

这里选择 Windows 7 版本的驱动程序，如图 4-14 所示。

（7）选择系统安装的磁盘

图 4-15 所示为选择系统安装的磁盘的界面。

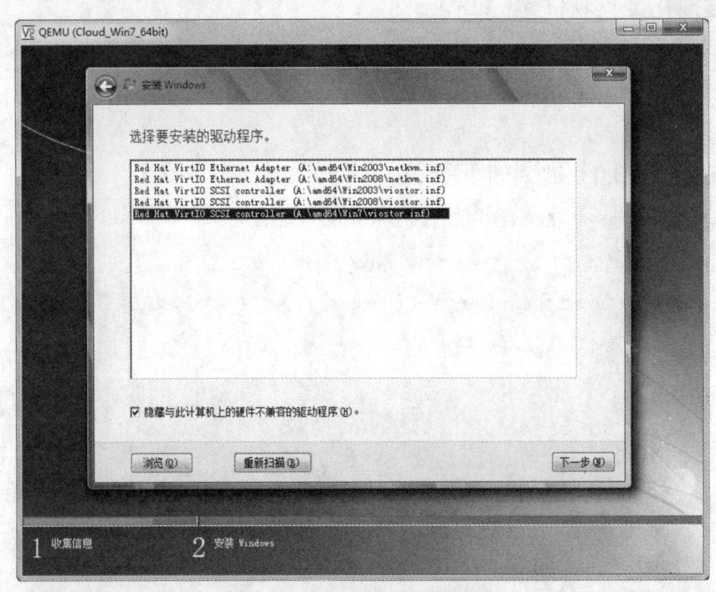

图 4-14　选择要安装的驱动

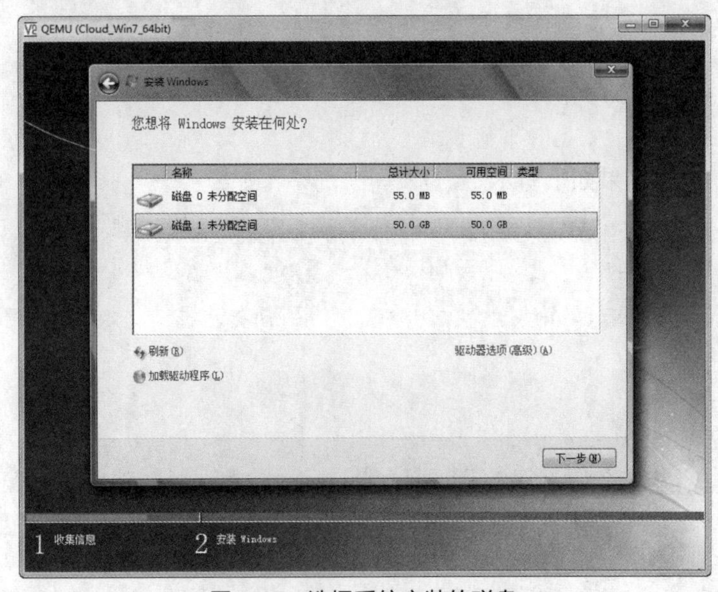

图 4-15　选择系统安装的磁盘

（8）开始安装

图 4-16 所示为开始安装的界面。

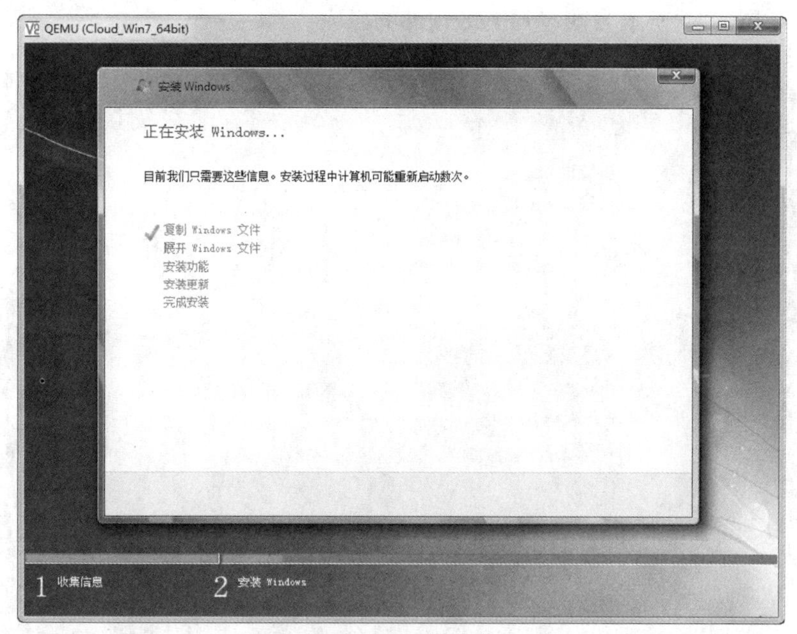

图 4-16　开始安装

（9）设置用户名密码

创建用户“CloudUser”，密码为“cloudpasswd”，如图 4-17 所示。

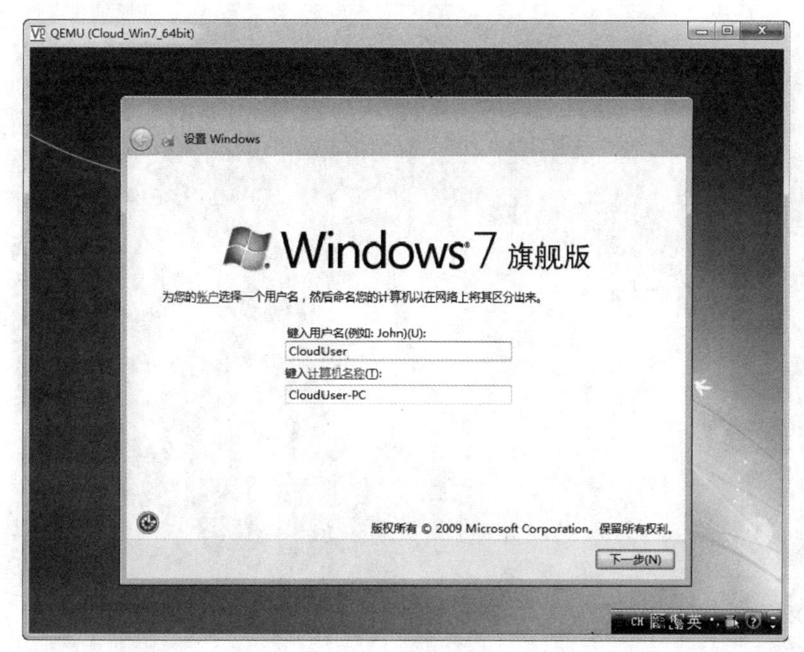

图 4-17　设置用户名

（10）登录系统

（11）下载并安装 Office 软件

图 4-18 和图 4-19 所示为安装 Office 软件的界面。

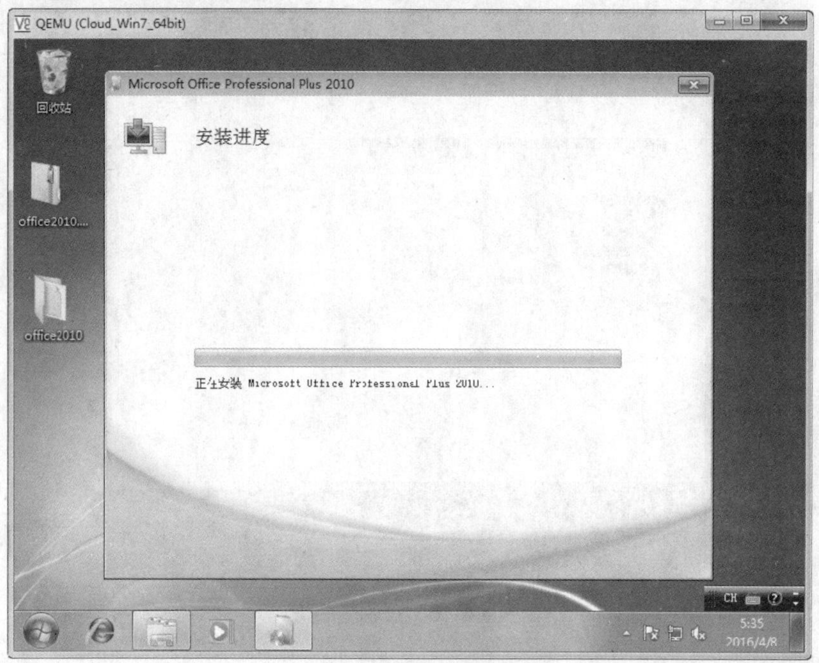

图 4-18　安装 Office 软件

图 4-19　Office 安装完成

（12）运行 JDK 安装文件，配置 Java 环境

图 4-20 所示为运行 JDK 安装文件的界面。

图 4-20　运行 JDK 安装文件

安装完 JDK 包后，需要设置环境变量。

打开“控制面板”→“系统安全”→“系统”→“高级系统设置”→“环境变量”。

首先新建系统变量，变量名为“JAVA_HOME”，变量值为“C:\Program Files\Java\jdk1.7.0_55”，即刚才 JDK 安装的路径，如图 4-21 所示。

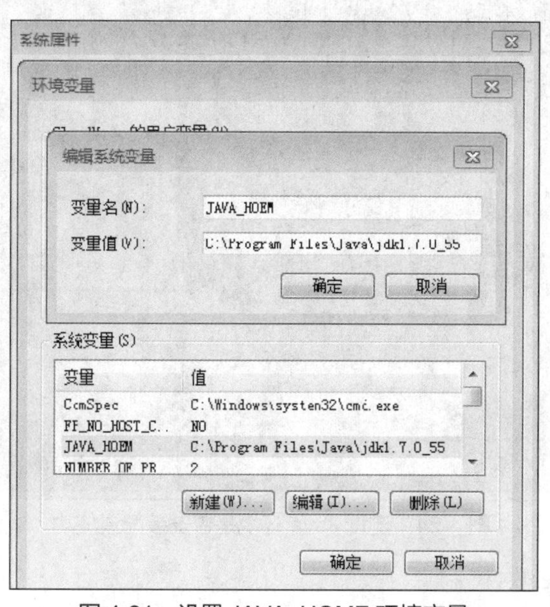

图 4-21　设置 JAVA_HOME 环境变量

为 CLASSPATH 变量添加新值“%JAVA_HOME%\lib\dt.jar;%JAVA_HOME%\lib\tools.jar;”，如图 4-22 所示。

为 PATH 变量添加新值“%JAVA_HOME%\bin;%JAVA_HOME%\jre\bin;”，如图 4-23 所示。

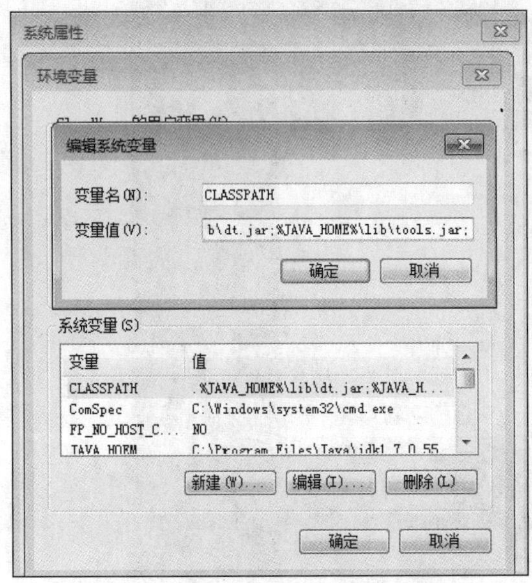

图 4-22 设置 CLASSPATH 环境变量

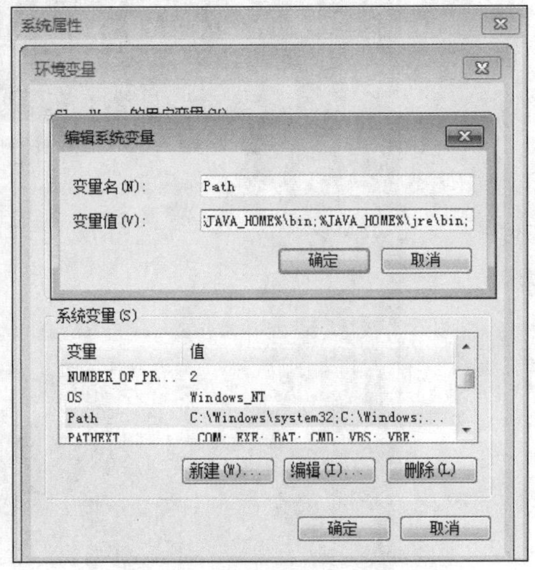

图 4-23 设置 PATH 环境变量

测试是否安装配置成功。打开运行工具，在命令提示符下输入“java –version”，弹出如下 Java 运行信息，如图 4-24 所示，说明安装配置成功。

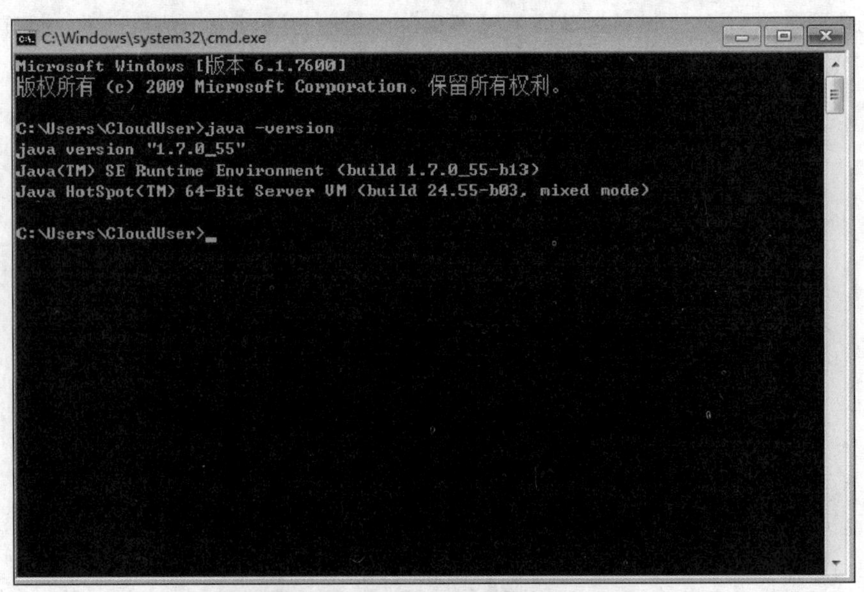

图 4-24 测试 Java 环境是否安装配置成功

（13）安装 Eclipse 软件

① 下载 Eclipse 软件包，运行 eclipse.exe 程序，如图 4-25 所示。

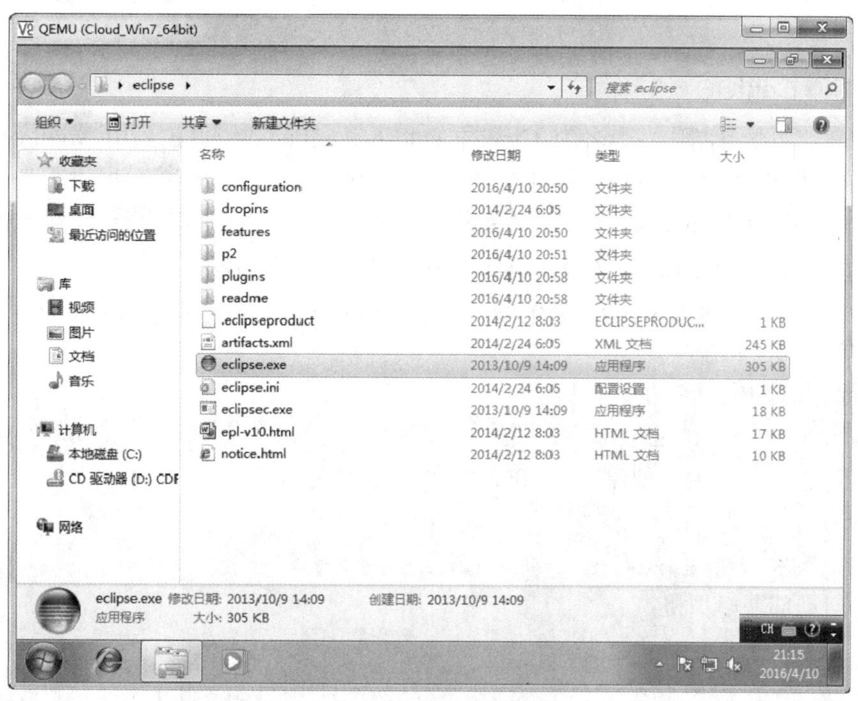

图 4-25　双击图标打开 Eclipse

② 选择工作目录，打开程序，Eclipse 工作界面如图 4-26 所示。

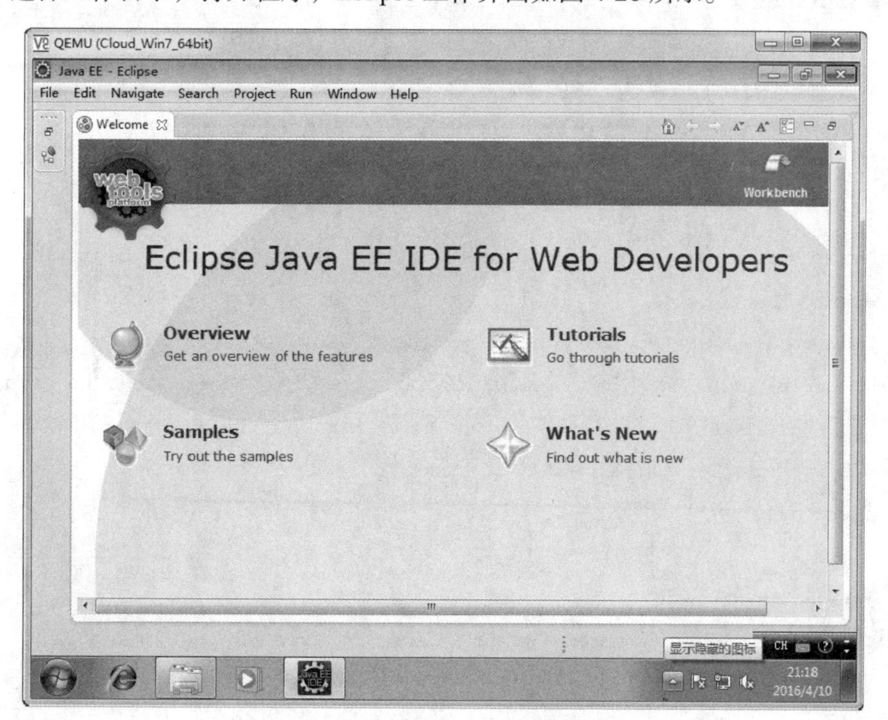

图 4-26　Eclipse 工作界面

（14）镜像格式转换

关闭虚拟机，开始转换镜像格式。

```
# qemu-img convert -f raw -O qcow2 Cloud_Win7_64bit.img  Cloud_Win7_ 64bit.qcow2
```

3. 制作 CentOS 6.5 镜像

接下来，小李要制作一个名为 Cloud_Centos6.5_64bit 的镜像，采用 Minimal Desktop 方式，root 根分区大小为 9GB，boot 分区大小为 200MB，swap 交换分区大小为 823MB，root 用户的密码为“cloudpasswd”，安装完成后检查 Java、Python 运行环境和 SSH 服务，预装 Eclipse，新建登录用户“CloudUser”，密码为“cloudpasswd”。

（1）首先要确保 CentOS-6.5-x86_64-bin_DVD.iso 镜像文件在服务器中，这里用户将该镜像文件放在/opt 目录中

查看镜像文件是否在指定位置。

```
# ll /opt/CentOS-6.5-x86_64-bin_DVD.iso
```

切换到/tmp 目录中，创建一个 10GB 大小的镜像文件。

```
# cd /tmp
# qemu-img create -f raw Cloud_Centos6.5_64bit.img 10G
```

（2）启动部署虚拟机

```
#  virt-install  --name  Cloud_Centos6.5_64bit  --ram  1024  --vcpus=1  --disk
path=/tmp/Cloud_Centos6.5_64bit.img  --network  network:default ， model=virtio
--arch=x86_64  --os-type=linux  --os-variant=rhel6  --graphics  vnc ， port=5910
--cdrom /opt/CentOS-6.5-x86_64-bin_DVD.iso --boot cdrom
```

执行结果如下所示。

```
[root@cl-controller~]# cd /tmp
[root@cl-controller tmp]# 11 /opt/CentOS-6.5-x86_64-bin DVD.iso
-rw-r--r--. 1 qemu qemu 4471971840 Mar 92015 /opt/CentOS-6.5-x86_64-bin DVD.i
so
[root@cl-controller tmp]# qemu-img create -f raw Cloud_Centos6.5_64bit·img 10G
Formatting  'Cloud Centos6.5_64bit.img', fmt=raw size=10737418240
[root@cl-controller tmp]# virt-install --name Cloud Centos6.5_ 64bit --ram 1024-
-vcpus=1 --disk path=/tmp/Cloud_Centos6.5_64bit.img --network network: default，mo
del-virtio --arch=x86_64 --os-type=linux --os-variant=rhel6 --graphics vnc，port=
5910 --cdrom /opt/CentOS-6.5-x86_64-bin DVD.iso --boot cdrom
Starting install…
Creating domain...|  0 B00:00
Cannot open display:
Run 'virt-viewer --help' to see a full list of available command line options
Domain installation still inprogress. You can reconnect to
the console to complete the installation process.
```

（3）通过 VNC Viewer 连接到虚拟桌面，安装虚拟机

（4）在对话框中输入 IP 端口号，连接到虚拟桌面

图 4-27 所示为 VNC 连接到虚拟桌面的界面。

（5）开始安装虚拟机操作系统

图 4-28 所示为开始安装 CentOS 操作系统。

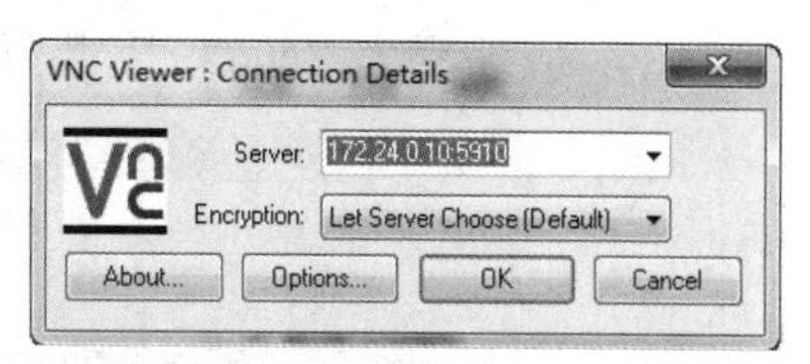

图 4-27　VNC 连接到虚拟桌面

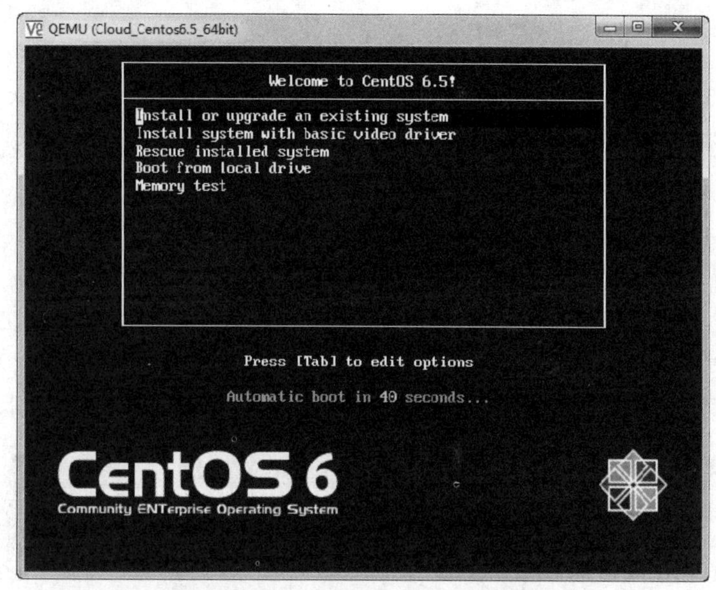

图 4-28　开始安装 CentOS 操作系统

（6）选择语言为英文

图 4-29 所示为选择操作系统语言的界面。

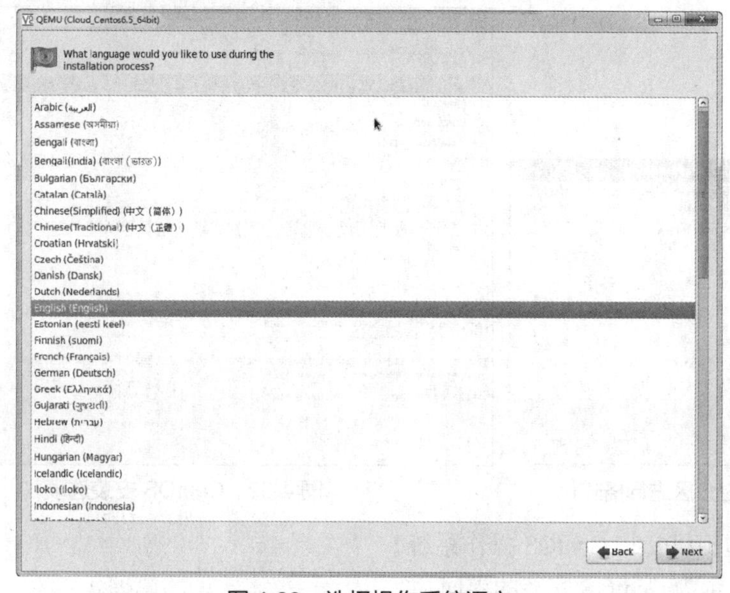

图 4-29　选择操作系统语言

（7）进行分区操作

图 4-30 所示为 CentOS 安装分区的界面。

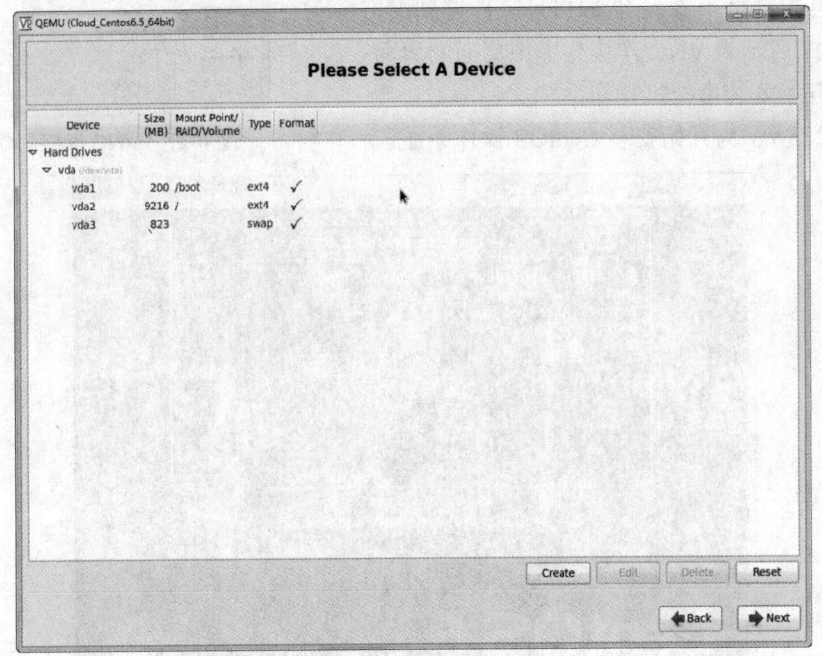

图 4-30　CentOS 安装分区

（8）进行格式化磁盘操作

图 4-31 所示为对分区进行格式化的界面。

（9）安装 Minimal Desktop 桌面版

图 4-32 所示为 CentOS 安装选项的界面。

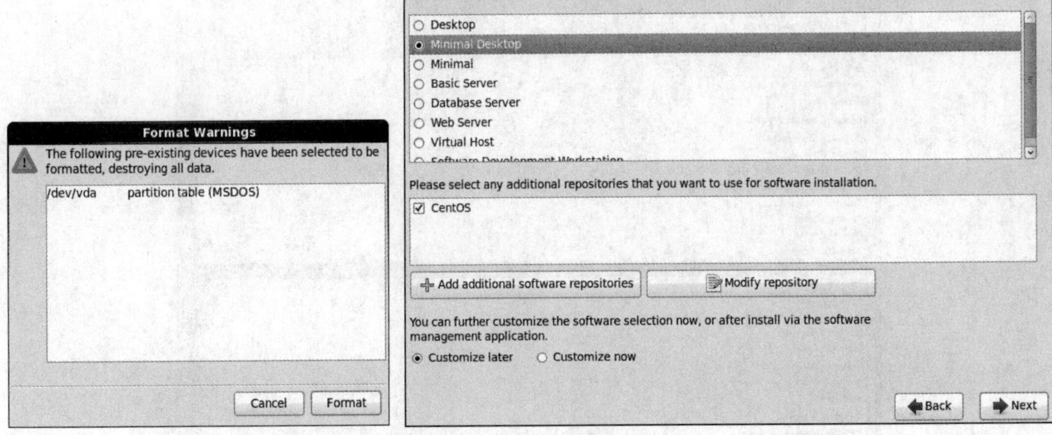

图 4-31　对分区进行格式化　　图 4-32　CentOS 安装选项

（10）为虚拟机安装 CentOS 操作系统

图 4-33 所示为 CentOS 安装的界面。

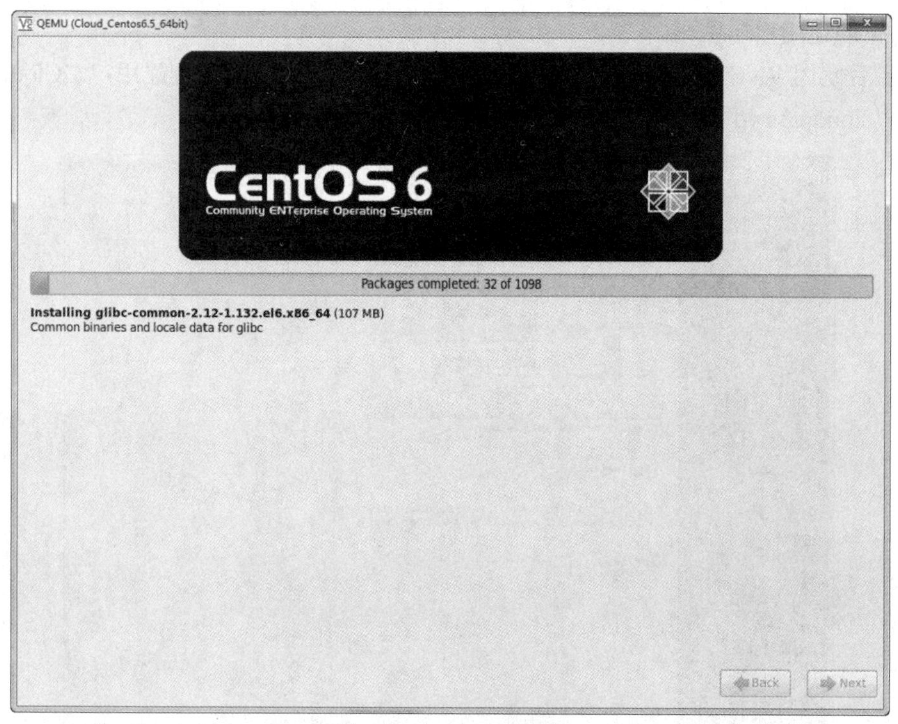

图 4-33 CentOS 安装

安装完成重启之后，查看虚拟机状态，启动虚拟机，重新通过 VNC Viewer 连接到虚拟桌面。

```
# virsh list -all
# virsh start Cloud_Centos6.5_64bit
# virsh list --all
```

执行结果如下所示。

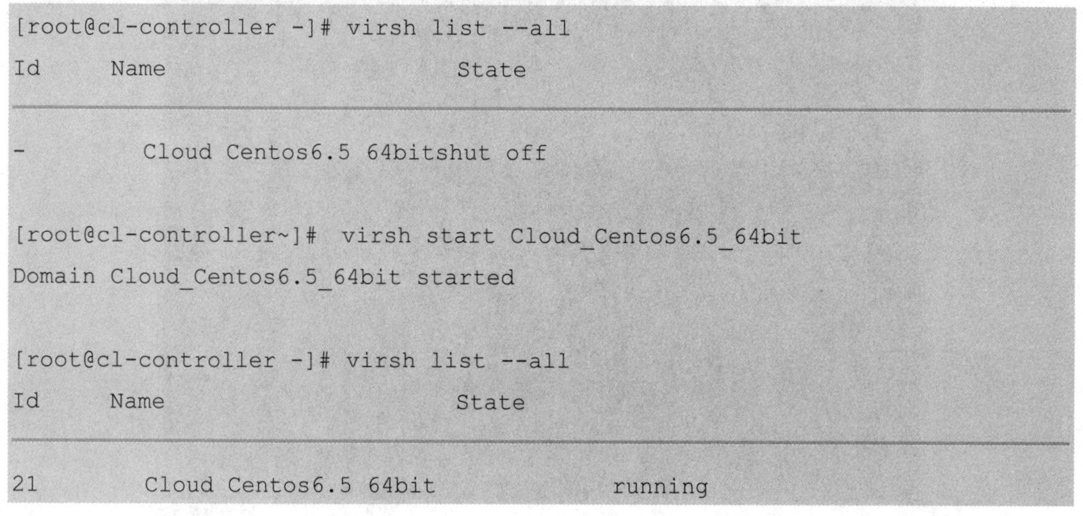

```
[root@cl-controller -]# virsh list --all
Id      Name                            State
---------------------------------------------------
-          Cloud Centos6.5 64bitshut off

[root@cl-controller~]#  virsh start Cloud_Centos6.5_64bit
Domain Cloud_Centos6.5_64bit started

[root@cl-controller -]# virsh list --all
Id      Name                            State
---------------------------------------------------
21         Cloud Centos6.5 64bit                running
```

（11）创建用户名密码

重启后使用 VNC 再次连接到虚拟机，在图 4-34 所示的界面中创建用户“CloudUser”，密码为“cloudpasswd”。

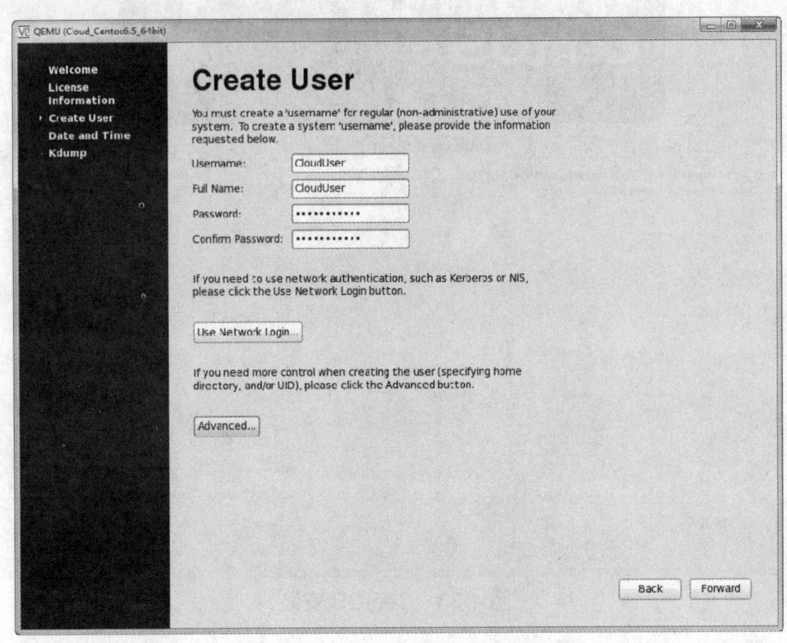

图 4-34　设置用户名密码

（12）登录安装好的操作系统

图 4-35 所示为进入虚拟机操作系统的界面。

图 4-35　进入虚拟机操作系统

（13）配置网络

命令如下所示。

```
DEVICE=eth0
HWADDR=52:54:00:70:8E:40
TYPE=Ethernet
UUID=15088830-e63e-46d8-9d47-a4a46affa3a9
ONBOOT=yes
NM_CONTROLLED=yes
BOOTPROTO=dhcp
```

在终端命令行提示符下编辑对应的配置文件，修改网络配置。

（14）重启网卡

命令如下所示。

```
[root@localhost ~]# vi /etc/sysconfig/network-scripts/ifcfg-eth0
[root@localhost ~]# service network restart
```

检查 Java、Python 运行环境和 SSH 服务，如下所示。

```
[root@localhost ~]# java -version
[root@localhost ~]# python -v
[root@localhost ~]#rpm - qa |grep ssh
```

（15）安装 Eclipse

执行如下脚本，安装 Eclipse。

```
# yum -y install eclipse
```

（16）为 Eclipse 创建快捷方式并运行

Eclipse 的路径为/usr/bin/eclipse，在桌面上单击鼠标右键创建快捷方式，如图 4-36 和图 4-37 所示。

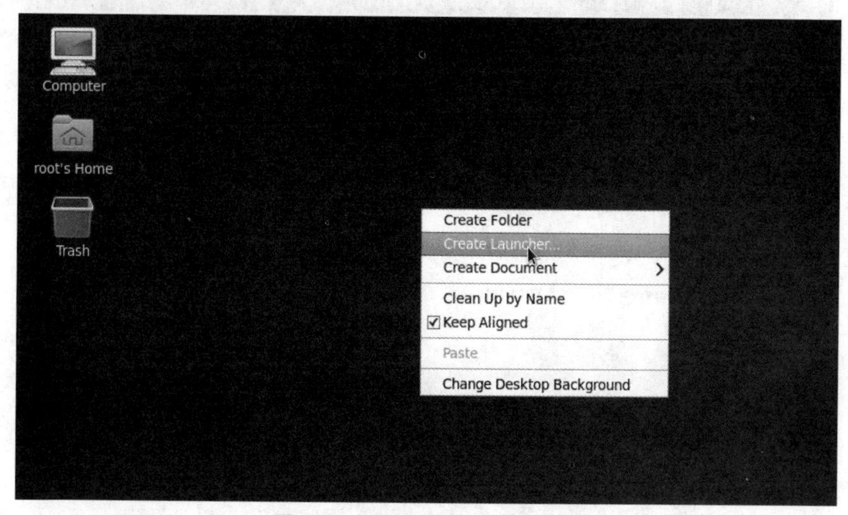

图 4-36　在桌面单击鼠标右键

图 4-37　创建快捷方式

快捷方式创建后，效果如图 4-38 所示。双击图标运行 Eclipse，效果如图 4-39 所示。

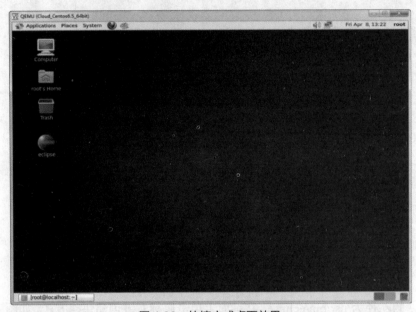

图 4-38　快捷方式桌面效果

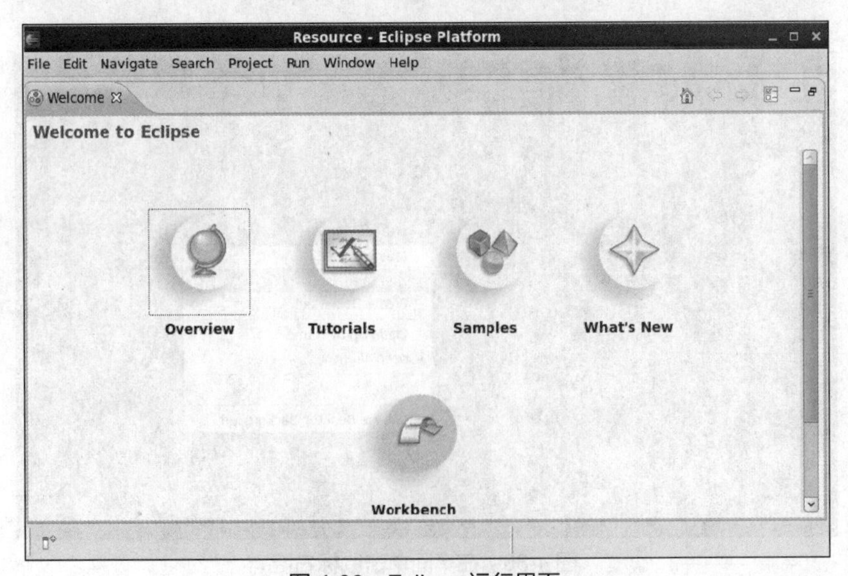

图 4-39　Eclipse 运行界面

（17）清除历史记录，关闭虚拟机

```
# history -c
# history -w
# shutdown -h now
```

（18）镜像格式转换

```
# qemu-img convert -f raw -O qcow2 Cloud_Centos6.5_64bit.img Cloud_Centos6.5_64bit.qcow2
```

4. 镜像上传

通过 GUI、CLI 方式上传镜像到系统内部，同时镜像为共有形式。（Windows 7 镜像上传名称为“Cloud_Win7_x64”；CentOS 镜像上传名称为“Cloud_Centos6.5_x64”）

通过 GUI 上传 Cloud_Win7_x64.qcow2 镜像的操作如图 4-40 所示。

图 4-40　上传镜像

结果如图 4-41 所示。

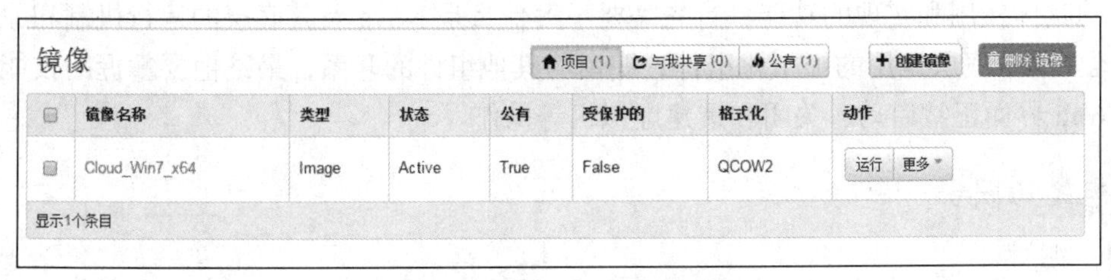

图 4-41　镜像管理的图形化界面

通过 CLI 上传 Cloud_Centos6.5_64bit.qcow2 镜像的命令如下。

```
# glance image-create --name Cloud_Centos6.5_64 --disk-format qcow2 --container-format bare --is-public true --progress < /tmp/Cloud_Centos6.5_ 64bit.qcow2
```

上传之后效果如下所示。

```
[root@controller ~]# glance image-create --name Cloud_Centos6.5_64 --disk-format qcow2 --container-format bare --is-public true --progress < /tmp/Cloud_Centos6.5_64bit.qcow2
[=============================>] 100%
```

Property	Value
checksum	71ff257b31ab72c1c7459146d57c7fb8
container_format	bare
created_at	2016-10-12T7:28:15
deleted	False
deleted_at	None
disk_format	qcow2
id	2d7e9141-7d18-45d4-91c6-d14d64bfb839
is_public	True
min_disk	0
min_ram	0
name	Cloud_Centos6.5_64
owner	219c95eac5694e45bd5c7304613835d3
protected	False
size	7446528000
status	active
updated_at	2016-10-12T7:28:15
virtual_size	None

任务三 学习计算服务

任务要求

小李现在已经大致了解到 Nova 服务对整个云计算平台的重要性，他还要深入了解 Nova 的具体作用和实现的功能，首先需要了解什么是 Nova 及其底层的运行机制和原理，其次他要了解底层调用的虚拟化组件，明白与其他组件的联系，最终他要掌握的技能是完成在 Web 界面正常启动、关闭和重建虚拟机等操作。

相关知识

1. 概述

计算服务（Nova）表示云平台的工作负载的核心。如果有些云服务的工作中不包括计算，那么它们充其量只代表静态存储，所有动态活动都会涉及一些计算元素。OpenStack Compute 这个名称指的是一个特定的项目，该项目也被称为 Nova，OpenStack 的其他组件依托 Nova，与 Nova 协同工作，组成了整个 OpenStack 云平台。

Nova 是 OpenStack 计算的弹性控制器。Nova 可以说是整个云平台最重要的组件，其功能包括运行虚拟机实例、管理网络以及通过用户和项目来控制对云的访问。OpenStack 最基础的开源项目名称为 Nova，它提供的软件可以控制基础设施即服务（IaaS）云计算平台，与 AmazonEC2 和 Rackspace 云服务器有一定程度的相似。OpenStack Compute 没有包含任何的虚拟化软件，相反，它定义和运行在主机操作系统上的虚拟化机制交互的驱动程序上，并通过基于 Web 的程序应用接口（API）来提供功能的使用。

2. 架构介绍

Nova 服务包含了 7 个子组件，分别为 Nova API、Nova Cert、Nova Compute、Nova Conductor、Nova Scheduler、Nova Consoleauth 以及 Nova Vncproxy。

（1）Nova API

Nova API 是一个 HTTP 服务，用于接收和处理客户端发送的 HTTP 请求。Nova API 是整个 Nova 组件的入口，它接受用户请求，将指令发送至消息队列（Queue），并由 Nova 服务的子服务执行相应的操作。

（2）Nova Cert

用于管理证书认证，提供兼容性保障，保证所有的应用程序都能在云上运行。

（3）Nova Compute

Nova Compute 是 Nova 的核心子组件，通过与 Nova Client 进行交互，实现了虚拟机管理的功能。它负责在计算节点上对虚拟机实例进行一系列操作，包括迁移、安全组策略和快照管理等功能。

（4）Nova Conductor

Nova Conductor 是 OpenStack 中的一个 RPC 服务，主要提供对数据库的查询和权限分配操作。

（5）Nova Scheduler

Nova Scheduler 可完成 Nova 的核心调度，包括虚拟机硬件资源调度、节点调度等。同时，Nova Scheduler 决定了虚拟机创建的具体位置。

（6）Nova Consoleauth

Nova Consoleauth 可实现对 Nova 控制台（VNC）的认证操作。

（7）Nova Vncproxy

Nova Vncproxy 是 Nova 的控制台组件，可实现客户端与虚拟机实例的通信，在虚拟桌面领域运用较广泛。

可以通过一个 Nova 实例的启动过程来了解内部的数据和请求的运行情况。用户首先会通过 Web 界面或者 CLI 界面发送一个启动实例的请求，需要通过认证服务 Keystone 的请求，进行身份的验证，然后将拿到的 Token 向 Nova API 发送请求，再通过 Glance 镜像服务检查镜像 image 和类型 flavor 是否存在。通过上述验证后将启动实例的请求发送给计算服务的 Schedule 调度节点，调度节点随机将此请求发送给一个 Compute 节点，让其启动实例，计算节点接受请求之后会到 Glance 存在的计算节点开始下载镜像并启动实例，计算节点启动虚拟机的时候会通过 Neutron 的 DHCP 获取一个对应的 IP 等网络资源，再次在 OVS 网桥获取相应的端口，绑定虚拟机的虚拟网卡接口。至此实例的启动已经完成。

Nova 集合了多个服务进程，每个进程扮演着不同的角色。用户通过 REST API 进行访问，Nova 的内部服务通过 RPC 消息进行服务通信。API 服务产生数据读写的 REST 请求，然后产生 REST 应答。内部服务通信通过 oslo.messaging 库的 RPC 消息实现消息的通信。大部分服务进程共用一个中心数据库，也可以把中心数据放到对象层，实现不同 OpenStack 发布版本的兼容性，如图 4-42 所示。

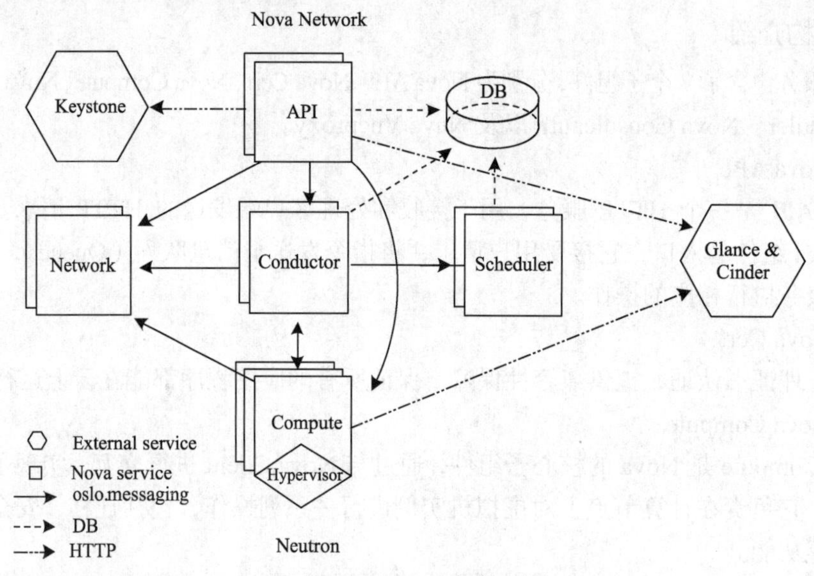

图 4-42　Nova 服务进程图

3. 调度机制（Scheduler）

Scheduler 模块在 OpenStack 中的作用就是决策虚拟机创建在哪个主机上，目前（截至 Essex 版本），调度仅支持计算节点。

（1）主机过滤

如图 4-43 所示，FilterScheduler 首先得到未经过滤的主机列表，然后根据过滤属性，选择主机创建指定数目的虚拟机。

目前，OpenStack 默认支持多种过滤策略，开发者也可以根据需要实现自己的过滤策略。在 nova.scheduler.filters 包中的过滤器有以下几种。

① AllHostsFilter：不做任何过滤，直接返回所有可用的主机列表。

② AvailabilityZoneFilter：返回创建虚拟机参数指定的集群内的主机。

③ ComputeFilter：根据创建虚拟机规格属性选择主机。

④ CoreFilter：根据 CPU 数过滤主机。

⑤ IsolatedHostsFilter：根据“image_isolated”和“host_isolated”标志选择主机。

⑥ JSONFilter：根据简单的 JSON 字符串指定的规则选择主机。

⑦ RAMFilter：根据指定的 RAM 值选择资源足够的主机。
⑧ SimpleCIDRAffinityFilter：选择在同一 IP 段内的主机。
⑨ DifferentHostFilter：选择与一组虚拟机不同位置的主机。
⑩ SameHostFilter：选择与一组虚拟机相同位置的主机。

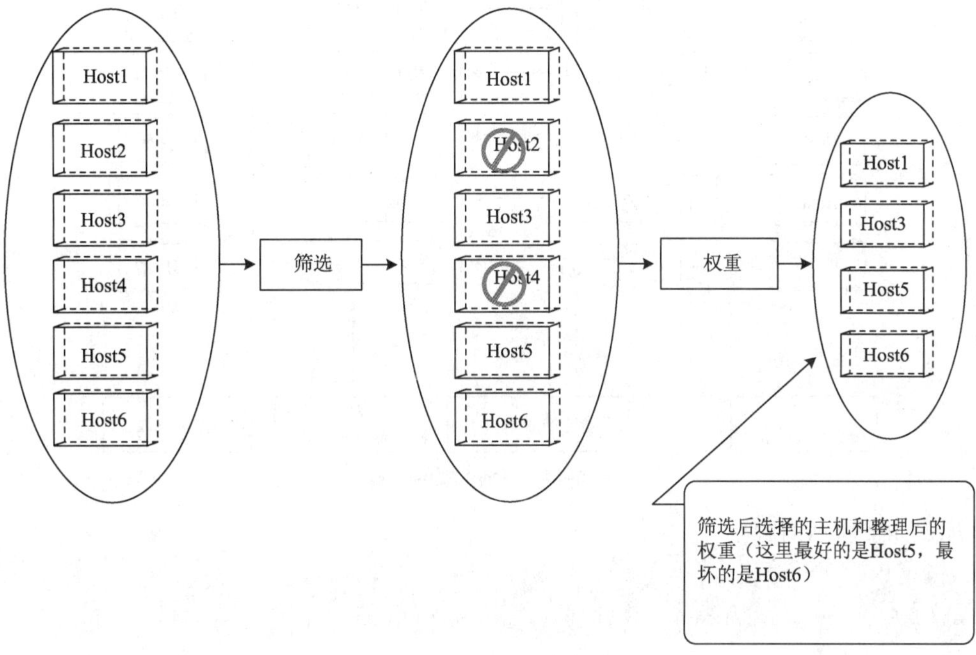

图 4-43　Nova 主机过滤

（2）权值计算

经过主机过滤后，需要对主机进行权值的计算，并根据策略选择相应的某一台主机（对于每一个要创建的虚拟机而言）。尝试在一台不适合的主机上创建虚拟机的代价比在一台合适的主机上创建的代价要高，例如，在一台高性能主机上创建一台功能简单的普通虚拟机的代价是高的。OpenStack 对权值的计算需要一个或多个（Weight 值，代价函数）的组合，然后对每一个经过过滤的主机调用代价函数进行计算，将得到的值与 Weight 值相乘，得到最终的权值。OpenStack 将在权值最小的主机上创建一台虚拟机，如图 4-44 所示。

（3）配置文件讲解

Nova 的安装与 Glance 组件的部署流程类似，Nova 的子服务更多，可把 Nova Compute 单独部署在计算节点，把 Nova 的其他服务部署在控制节点。

为了实现计算节点的 Nova Compute 服务与控制节点上 Nova 其他的子服务通信，需要在配置文件中配置 QPID 消息队列服务。

```
[root@controller ~]# vi /etc/nova/nova.conf
```

配置如下所示内容。

```
[DEFAULT]
rpc_backend = qpid
qpid_hostname = controller
```

Nova 服务的数据存储在 MySQL 数据库中，需要在数据库中创建 Nova 数据库并给予一定的权限，并在 Nova 的配置文件中配置数据库链接，指向 MySQL 中的 Nova 数据库。

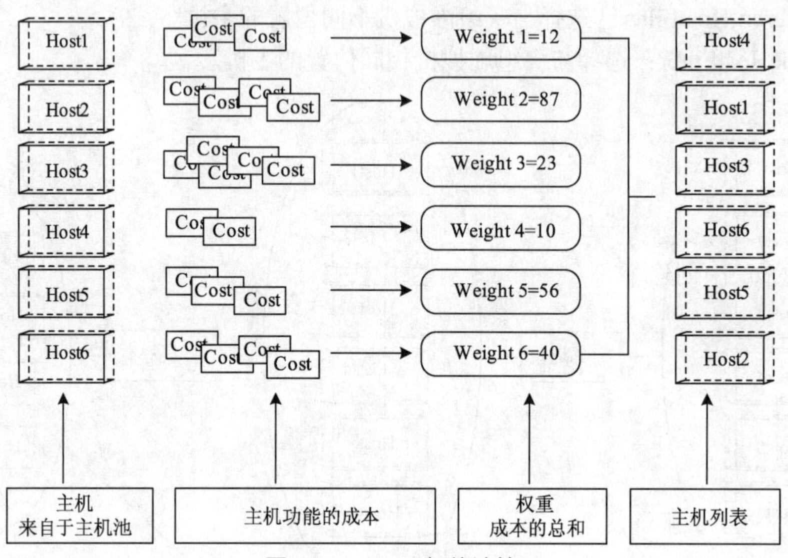

图 4-44　Nova 权值计算

配置如下所示内容。

```
[database]
connection = mysql://nova:000000@controller/nova
```

```
#mysql -uroot -p
```

执行如下所示的命令，赋予数据库权限。

```
mysql> CREATE DATABASE nova;
mysql> GRANT ALL PRIVILEGES ON nova.* TO 'nova'@'localhost' IDENTIFIEDBY 'NOVA_
DBPASS';
mysql> GRANT ALL PRIVILEGES ON nova.* TO 'nova'@'%' IDENTIFIEDBY 'NOVA_DBPASS';
```

```
# su -s /bin/sh -c "nova-manage db sync" nova
```

与 Glance 组件一样，需要在 Keystone 创建 Nova 用户，执行的命令和结果如下所示。

```
# keystone user-create --name=nova --pass=NOVA_PASS --email=NOVA_EMAIL
```

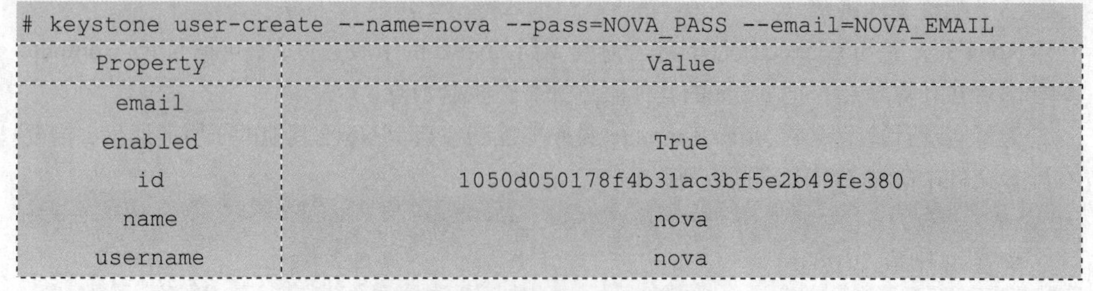

Property	Value
email	
enabled	True
id	1050d050178f4b31ac3bf5e2b49fe380
name	nova
username	nova

赋予 Nova 用户"admin"权限，并把 Nova 用户规划到"service"租户下，同时在 Keystone 中注册 Nova 服务，服务类型为 Compute。执行的命令和结果如下所示。

```
# keystone user-role-add --user=nova --tenant=service --role=admin
```

```
# keystone service-create --name=nova --type=compute --description= "OpenStack Compute"
```

Property	Value
description	OpenStack Compute
enabled	True
id	ba7d105e1dd346cab67334796869a00c
name	nova
type	compute

在 Keystone 中创建 Nova 计算服务的端点。执行的命令和结果如下所示。

```
# keystone endpoint-create --service-id=$(keystone service-list | awk '/ compute / {print
$2}') --publicurl=http://controller:8774/v2/%\(tenant_id\)s
--internalurl=http://controller:8774/v2/%\(tenant_id\)s
--adminurl=http://controller:8774/v2/%\(tenant_id\)s
```

为了实现 Nova 服务调用其他服务，需要在 Nova 的配置文件中配置其通过 Keystone 的认证。

```
# vi /etc/nova/nova.conf
[keystone_authtoken]
auth_uri = http://controller:5000
auth_host = controller
auth_protocol = http
auth_port = 35357
admin_user = nova
admin_tenant_name = service
admin_password = 000000
```

为了实现对平台虚拟机的访问，OpenStack 提供了 VNC 的接入方式，需要在 Nova 的配置文件中配置 VNC 接入。

```
[DEFAULT]
rpc_backend = qpid
qpid_hostname = controller
my_ip = 172.24.2.10
vncserver_listen = 172.24.2.10
vncserver_proxyclient_address = 172.24.2.10
```

（4）Nova 的主要操作命令

Nova 作为 OpenStack 的核心组件，拥有强大的功能、权限，可以对整个平台的资源（镜像、网络和存储等）进行管理，下面来看一下 Nova 的主要操作命令。

① Nova 对镜像进行管理。

- nova image-create

功能：通过快照创建镜像。

格式：

```
# nova image-create [--show] [--poll] <server><name>
```

固定参数说明如下所示。

<server>：做快照的实例的名字或者 ID。

<name>：快照名字。

可选参数说明如下所示。

[--show] ：打印出镜像信息。

[--poll]：显示进度。

- nova image-show

功能：获取镜像的详细信息。

格式：nova image-show <image>

参数说明如下所示。

<image>：所要查询的镜像的名字或者 ID。

```
# nova image-list
```

执行结果如下所示。

ID	Name	Disk Format	Container Format	Size	Status
f7cac855-e4a1-4690-bd1b-f8d79f90286f	centos6.5	qcow2	bare	305397760	active
5d93b088-0fae-4f3e-af83-5eb4195226f7	cirros-0.3.4-x86_64	qcow2	bare	13287936	active

```
# nova image-show f7cac855-e4a1-4690-bd1b-f8d79f90286f
```

具体执行结果如下所示。

Property	Value
OS-EXT-IMG-SIZE:size	305397760
created	2016-10-11T06:20:49
id	f7cac855-e4a1-4690-bd1b-f8d79f90286f
min_disk	0
min_ram	0
name	centos6.5
progress	100
status	ACTIVE
updated	2016-10-11T06:20:51

② Nova 管理安全组规则。

安全组，翻译成英文是 Security Group。安全组是一些规则的集合，用来对虚拟机的访问流量加以限制，这反映到底层，就是使用 IPTables 给虚拟机所在的宿主机添加 IPTables 规则。可以定义 *n* 个安全组，每个安全组可以有 *n* 个规则，可以给每个实例绑定 *n* 个安全组，Nova 中总是有一个 default 安全组，这个是不能被删除的。创建实例的时候，如果不指定安全组，会默认使用这个 default 安全组。目前，Nova 中的安全组只是对进入虚拟机的流量加以控制，对虚拟机外出流量没有加以限制。Nova 中安全组应该会移到 Neutron 中，并且会增加对虚拟机外出流量的控制。

常用的安全组命令如下所示。

```
# nova secgroup-create
```

功能：创建安全组。

格式：

```
# usage: nova secgroup-create <name><description>
```

参数说明如下所示。

<name>：安全组名字。

<description>：安全组描述。

举例：

```
# nova secgroup-create test 'test the nova command about the rules'
```

Id	Name	Description
249c535a-c858-404a-8900-2bfe3d752c6b	test	test the nova command about the rules

```
# nova secgroup-add-rule
```

功能：给安全组添加规则。

格式：

```
usage: nova secgroup-add-rule <secgroup><ip-proto><from-port><to-port>
<cidr>
```

参数说明如下。

<secgroup>：安全组名字或者 ID。

<ip-proto> IP：协议(ICMP，TCP，UDP)。

<from-port>：起始端口。

<to-port>：结束端口。

<cidr>：网络地址。

具体执行结果如下所示。

```
# nova secgroup-add-rule test icmp -1 -1 0.0.0.0/0
```

IP Protocol	From Port	To Port	IP Range	Source Group
Icmp	-1	-1	0.0.0.0/0	

（5）Nova 虚拟机类型管理

虚拟机类型是在创建实例的时候分配给实例的资源情况，下面来看一下 Nova 对虚拟机类型的管理。

```
# nova flavor-create
```

功能：创建一个虚拟机类型。

格式：

```
# nova flavor-create          [--ephemeral <ephemeral>] [--swap <swap>]
                              [--rxtx-factor <factor>] [--is-public<is-public>]
```

```
<name><id><ram><disk><vcpus>
```

固定参数说明如下所示。

<name>：虚拟机类型名字。

<id>：虚拟机类型 ID。

<ram>：虚拟内存大小。

<disk>：磁盘大小。

<vcpus>：虚拟内核个数。

可选参数说明如下所示。

[--ephemeral <ephemeral>]：临时磁盘大小，单位为 GB。

[--swap <swap>]：交换分区大小。

[--rxtx-factor <factor>]：传输因子，默认为 1。

[--is-public <is-public>]：是否共享。

具体执行结果如下所示。

```
# nova flavor-create test 3 2048 20 2
```

ID	Name	Memory_MB	Disk	Ephemeral	Swap	VCPUs	RXTX_Factor	Is_Public
3	test	2048	20	0		2	1.0	True

```
# nova flavor-show test
```

Property	Value
OS-FLV-DISABLED:disabled	False
OS-FLV-EXT-DATA:ephemeral	0
disk	20
extra_specs	{}
id	3
name	Test
os-flavor-access:is_public	True
ram	2048
rxtx_factor	1.0
swap	
vcpus	2

（6）Nova 实例管理

Nova 可对云平台中的实例进行管理，包括创建实例、启动实例、删除实例和实例迁移等操作。

```
# nova boot
```

功能：启动实例。

格式：

```
#  nova boot [--flavor <flavor>] [--image <image>]
              [--image-with <key=value>] [--boot-volume <volume_id>]
              [--snapshot <snapshot_id>] [--num-instances <number>]
```

```
              [--meta<key=value>] [--file <dst-path=src-path>]
              [--key-name <key-name>] [--user-data <user-data>]
              [--availability-zone <availability-zone>]
              [--security-groups <security-groups>]
              [--block-device-mapping <dev-name=mapping>]
              [--block-device key1=value1[, key2=value2...]]
              [--swap <swap_size>]
              [--ephemeral size=<size>[, format=<format>]]
              [--hint <key=value>]
              [--nic <net-id=net-uuid, v4-fixed-ip=ip-addr, port-id=port-uuid>]
              [--config-drive <value>] [--poll]
<name>
```

固定参数说明如下所示。

<name>：实例名称。

可选参数说明如下所示。

[--flavor <flavor>]：虚拟机类型。

[--image <image>]：选用的镜像。

[--image-with <key=value>] ：镜像的元数据属性。

[--boot-volume <volume_id>：启动逻辑卷的 ID。

[--snapshot <snapshot_id>]：快照。

[--num-instances <number>]：实例数量。

[--meta <key=value>]：元数据。

[--file <dst-path=src-path>]：文件。

[--key-name <key-name>]：密钥名称。

[--user-data <user-data>]：注入的用户数据。

[--availability-zone <availability-zone>]：可用域。

[--security-groups <security-groups>]：安全组。

[--block-device-mapping <dev-name=mapping>]：块存储格式化。

[--block-device key1=value1[，key2=value2...]]：块设备参数。

[--swap <swap_size>]：交换分区大小。

[--ephemeral size=<size>[，format=<format>]]：连接块存储大小。

[--hint <key=value>]：自定义数据。

[--nic]：配置 IP。

[--config-drive <value>]：驱动使能。

[--poll]：显示创建进度。

具体执行结果如下所示。

```
# nova boot --flavor 3 --nic net-id=019c863c-9ce1-40b5-b2fb-ef014fe7c4ec--image
f7cac855-e4a1-4690-bd1b-f8d79f90286f test
+------------------------------------------------+------------------------------+
| Property                                       | Value                        |
+------------------------------------------------+------------------------------+
```

OS-DCF:diskConfig	MANUAL
OS-EXT-AZ:availability_zone	nova
OS-EXT-SRV-ATTR:host	-
OS-EXT-SRV-ATTR:hypervisor_hostname	-
OS-EXT-SRV-ATTR:instance_name	instance-00000004
OS-EXT-STS:power_state	0
OS-EXT-STS:task_state	Scheduling
OS-EXT-STS:vm_state	Building
OS-SRV-USG:launched_at	-
OS-SRV-USG:terminated_at	-
accesslPv4	
accesslPv6	
adminPass	4iKw7NQPxR5U
config_drive	
created	2016-08-11T00:48:432
flavor	test (3)
hostld	
id	384d6c42-56f1-4ee4-8dfe-4e8d413a0793
image	centos6.5 (f7cac855-e4al-4690-bdlb-f8d79f90286f)
key_name	-
metadata	{}
name	test
os-extended-volumes: volumes_attached	[]
progress	0
security_groups	default
status	BUILD
tenant_id	a3c430debe7e48d4a2f58c96dd89746b
updated	2016-08-11T00:48:43Z
user_id	db21d0b623e84e48bb1f92b281dcbfa3

```
# nova list
```

ID	Name	Status	Power State	Networks
b46418fb-106f-4438-8b5a-567fd967c861	Cloud test	ACTIVE	Running	int-net=172.24.4.2, 172.24.3.7
384d6c42-56f1-4ee4-8dfe-4e8d413a0793	test	ACTIVE	Running	int-net=172.24.4.4

（7）Nova 对逻辑卷的管理

Nova 对逻辑卷的管理也是 Nova 的一个比较重要的功能，下面来看一下 Nova 对逻辑卷管理的常用命令。

```
# nova volume-create
```

功能：通过 Nova 创建逻辑卷。

格式：

```
# nova volume-create [--snapshot-id <snapshot-id>]
                     [--image-id <image-id>]
                     [--display-name <display-name>]
                        [--display-description
                            <display-description>]
                     [--volume-type <volume-type>]
                     [--availability-zone
<availability-zone>]
<size>
```

固定参数说明如下所示。

<size>：大小。

可选参数说明如下所示。

[--snapshot-id <snapshot-id>]：通过快照创建逻辑卷的快照 ID。

[--image-id <image-id>]：通过镜像创建逻辑卷的镜像 ID。

[--display-name <display-name>]：逻辑卷名称。

[--display-description <display-description>]：逻辑卷描述。

[--volume-type <volume-type>：逻辑卷类型。

[--availability-zone <availability-zone>]：逻辑卷可用域。

创建一个名字为“test”、大小为 2GB 的云硬盘，命令如下所示。

```
# nova volume-create --display-name test 2
```

具体执行结果如下所示。

Property	Value
attachments	[]
availability_zone	nova
bootable	false
created_at	2016-08-11T01:09:06.601336
display_description	-
display_name	test
encrypted	False
id	793bc0e6-62e6-463a-b15d-24a7a13cf420
metadata	{}
size	2
snapshot_id	-
sourcevolid	-
status	creating
volume_type	None

（8）Nova 的其他命令

```
# nova bash-completion
restore --boot-volume floating-ip-disassociate --port help secgroup-delete-rule
--priority --ipv6 --file --image-with server-group-list --instance-name suspend
--action  --login  flavor-access-list  --dns1  --display-description  host-meta
--end floating-ip-list --swap --pm_user pause --ephemeral --snapshot secgroup-
update cell-capacities get-vnc-console -i -h secgroup-create  ......
```

（9）拓展延伸

① Nova 组件的调用实现。

OpenStack 的各个服务之间通过统一的 REST 风格的 API 调用，实现系统的松耦合。松耦合架构的好处是，各个组件的开发人员可以只关注各自的领域，对各自领域的修改不会影响到其他开发人员。不过，从另一方面来讲，这种松耦合的架构也给整个系统的维护带来了一定的困难，运维人员要掌握更多的与系统相关的知识去调试出了问题的组件。所以，无论对于开发还是维护人员，搞清楚各个组件之间的相互调用关系都是非常必要的，用户可以从 Nova Client 入手来了解其中的过程。

Nova Client 是一个命令行的客户端应用，终端用户可以从 Nova Client 发起一个 API 请求到 Nova API，Nova API 服务会转发该请求到相应的组件上。同时，Nova API 支持对 Cinder 和 Neutron 的请求转发，也就是可以在 Nova Client 上直接向 Cinder 和 Neutron 发送请求。用户可以在调用 Nova Client 时增加“--debug”选项，以打印更多的 debug 信息，通过这些 debug 信息可以了解到，用户需要发起一个完整的业务层面上的请求，都需要与哪些服务打交道。以启动一个实例的过程举例，如下。

```
#  nova --debug boot  --flavor 3 \
--nic net-id=54cf9939-4d75-4ba2-9b27-6becf8f13561 \
--image cfaf8992-bbb9-4895-bda1-a1547f0fa356 test1
```

输出结果详情见附录八 nova–debug.txt。

用户可以清楚看到，执行一个 boot 新实例的操作需要发送如下几个 API 请求。

- 向 Keystone 发送请求，获取租户（d7beb7f28e0b4f41901215000339361d）的认证 Token。
- 通过拿到的 Token，向 Nova API 服务发送请求，验证 image 是否存在。
- 通过拿到的 Token，向 Nova API 服务发送请求，验证创建的 favor 是否存在。
- 请求创建一个新的 instance，需要的元数据信息包含在请求 body 中。

Nova Client 可帮助用户把需要的全部请求放到一起，而最重要的就是如上所示的第 4 种请求。如果用户想自己通过 REST API 直接发送 HTTP 请求的话，可以直接使用第 4 种请求，当然，前提是先通过调用 Keystone 服务得到认证 Token。

下面结合代码重点叙述一下在第 4 种情况下的请求数据流动在整个 Open Stack 中的过程。图 4-45 是一个全局的流程图，图中每个服务都是一个单独的进程实例，它们之间通过 RPC 进行调用（或者广播）。

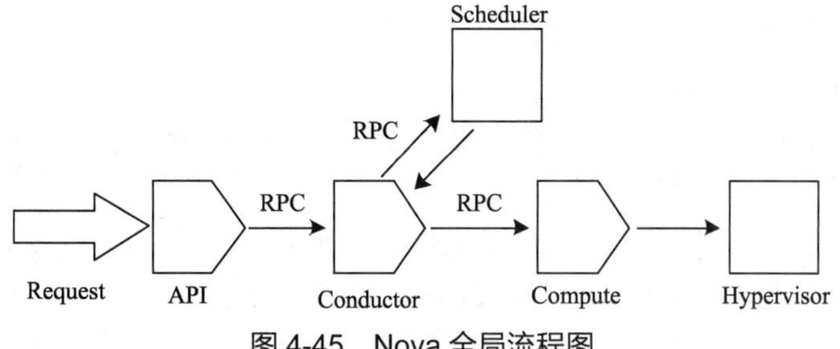

图 4-45　Nova 全局流程图

以下是上面涉及的服务的主要功能。

- Nova API：接受 HTTP 请求，并响应请求，当然还包括对请求信息的验证。
- Nova Conductor：与数据库交互，提高对数据库访问的安全性。
- Nova Scheduler：调度服务，决定最终实例要在哪个服务上创建。迁移、重建等都需要通过这个服务。
- Nova Compute：调用虚拟机管理程序，完成虚拟机的创建、运行以及控制。

② 数据库结构讲解。

跟其他组件相似，Nova 也把数据存储在 MySQL 中，Nova 的每一步操作都记录在数据库中，实例、网络、用户和租户通过关系型数据库相互依赖。

③ KVM、Libvirt 与 OpenStack。

1959 年 6 月，Christopher Strachey 发表虚拟化论文，虚拟化是今天云计算基础架构的基石。KVM 是最底层的 Hypervisor，它是用来模拟 CPU 运行的，它缺少了对 network 和周边 I/O 的支持，所以用户是没法直接使用它的。QEMU-KVM 就是一个完整的模拟器，它是构建于 KVM 之上的，提供了完整的网络和 I/O 支持。OpenStack 不会直接控制 QEMU-KVM，它会用一个称为 Libvit 的库去间接控制 QEMU-KVM，Libvirt 提供了跨 VM 平台的功能，它可以控制除 QEMU 外的模拟器，包括 VMware、VirtualBox、Xen 等。所以，为了 OpenStack 的跨 VM 性，OpenStack 只用 Libvirt 而不直接用 QEMU-KVM，如图 4-46 所示。Libvirt 还提供了一些高级的功能，如 pool/vol 管理。

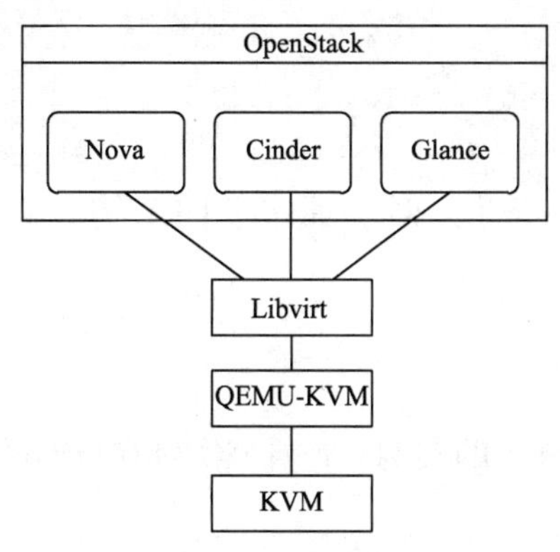

图 4-46　OpenStack 与 KVM

总结如下。

- Libvirt 是一套用 C 语言写的 API，旨在为各种虚拟机提供一套通用的编程接口，而且支持与 Java、Python 等语言的绑定。

- Libvirt 由几个不同的部分组成，其中包括应用程序编程接口（API）库、一个守护进程（Libvirtd），以及一个默认命令行实用工具（Virsh），Libvirtd 守护进程负责对虚拟机的管理工作。
- KVM 负责 CPU 虚拟化和内存虚拟化，实现了 CPU 和内存的虚拟化，但 KVM 不能模拟其他设备。Libvirt 调用 KVM 虚拟化技术的接口是用于管理的，用 Libvirt 管理方便。

④ API 解析。

在前面用户已经接触到了 API，API 是外部程序入口，OpenStack 作为开源的平台，开发人员正是通过 OpenStack 每个组件的 API 实现组件服务调用和二次开发的。用户简单地执行的一条关于 Nova 的命令，效果如下所示。

```
# nova list
```

ID	Name	Status	Power State	Networks
b46418fb-106f-4438-8b5a-567fd967c861	Cloud test	ACTIVE	Running	int-net=172.24.4.2, 172.24.3.7
384d6c42-56f1-4ee4-8dfe-4e8d413a0793	test	ACTIVE	Running	int-net=172.24.4.4

这条命令究竟经过了哪些过程呢?

具体来说，执行 Nova List 的过程就是经历了两次 HTTP 请求。

⑤ Keystone 的认证。

```
curl -i 'http://172.24.0.10:35357/v2.0/tokens' -X POST -H "Content-Type: application/json" -H "Accept: application/json" -H "User-Agent: python-novaCLIent" -d '{"auth": {"tenantName": "admin", "passwordCredentials": {"username": "admin", "password": "000000"}}}'
```

⑥ 调用 Nova API 取得所有实例。

```
curl -i 'http://172.24.2.10:8774/v2/219c95eac5694e45bd5c7304613835d3/ servers/detail' -X GET -H "X-Auth-Project-Id: admin" -H "User-Agent: python- novaCLIent" -H "Accept: application/json" -H "X-Auth-Token: $token"
```

从上面的结果中看到了一条条 URL，而这些 URL 构成了整个 OpenStack API 的基础。

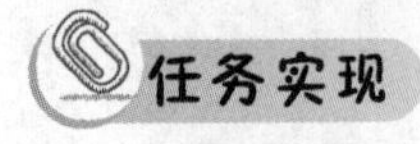

启动实例

启动一个名为“Cloud_test”的实例，查看该实例在机器的具体位置，查看其具体的配置信息和模板信息。

通过 GUI 界面创建一个 Cloud_test 的实例。

① 进入 Dashboard 界面找到“项目”选项。

② 打开 compute 面板，选择实例。

③ 选择启动云主机。

④ 在弹出的窗口中输入“云主机名称”，选择“云主机类型”，设置“云主机启动源”为“从镜像启动”，选择“镜像名称”，如图 4-47 所示。

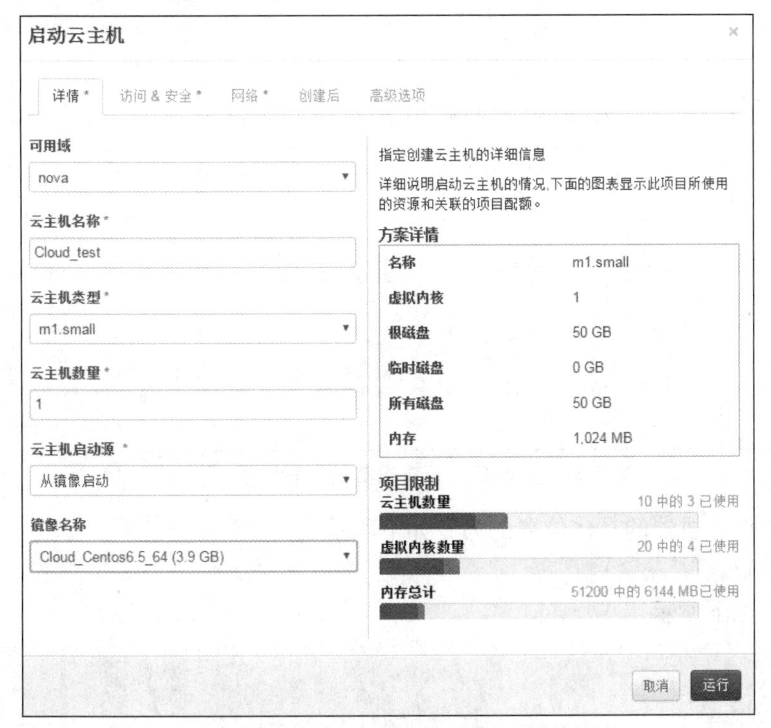

图 4-47　通过 GUI 界面创建一个实例

⑤ 在网络窗口中选择云主机的所属网络，如图 4-48 所示。

图 4-48　选择云主机的所属网络

⑥ 在 Dashboard 界面中查看实例，如图 4-49 所示。

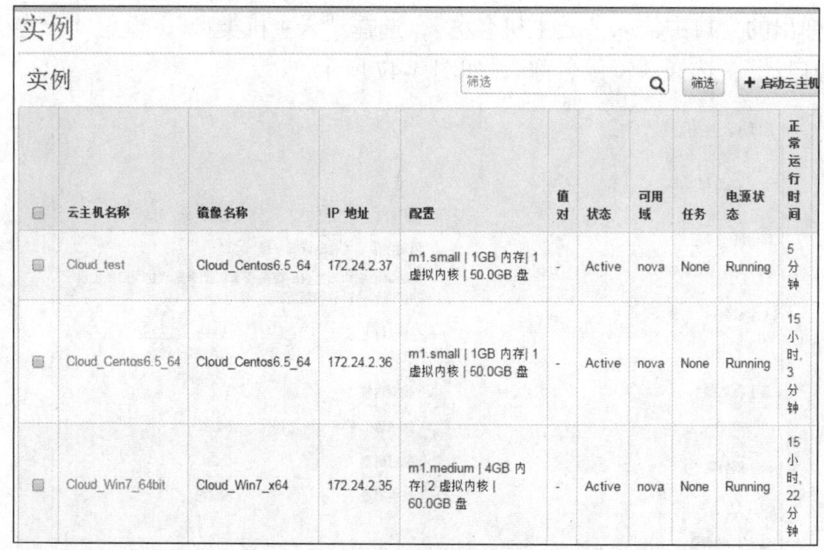

实例

实例 筛选 筛选 + 启动云主机

	云主机名称	镜像名称	IP 地址	配置	值对	状态	可用域	任务	电源状态	正常运行时间
	Cloud_test	Cloud_Centos6.5_64	172.24.2.37	m1.small | 1GB 内存| 1 虚拟内核 | 50.0GB 盘	-	Active	nova	None	Running	5 分钟
	Cloud_Centos6.5_64	Cloud_Centos6.5_64	172.24.2.36	m1.small | 1GB 内存| 1 虚拟内核 | 50.0GB 盘	-	Active	nova	None	Running	15 小时，3 分钟
	Cloud_Win7_64bit	Cloud_Win7_x64	172.24.2.35	m1.medium | 4GB 内存| 2 虚拟内核 | 60.0GB 盘	-	Active	nova	None	Running	15 小时，22 分钟

图 4-49 云主机实例列表界面

查看该实例在机器的具体位置的命令如下所示。

```
# nova show Cloud_test
```

执行结果如下所示。

Property	Value
OS-DCF:diskConfig	AUTO
OS-EXT-AZ:availability_zone	nova
OS-EXT-SRV-ATTR:host	compute
OS-EXT-SRV-ATTR:hypervisor_h	compute
ostname	instance-00000003
OS-EXT-SRV-ATTR:instance_name	1
OS-EXT-STS:power_state	-
OS-EXT-STS:task_state	active
OS-EXT-STS:vm_state	2016-08-10T06:35:21.000000
OS-SRV-USG:launched_at	-
OS-SRV-USG:terminated_at	
accesslPv4	
accesslPv6	
config_drive	2016-08-10T06:35:10Z
created	m1.5mall (c7bf1a9c-fdd3-f71f-af10-ca34c9402320
flavor	ddee54eeba959e7b72f50767ce935531a1ef6f1c29
hostld	fcb6933bf88646
id	b46418fb-106f-4438-8b5a-567fd967c861
image	Cloud_Cento56.5_64
int-net network	(ad019f85-f93a-4c6c-b7b0-e474e2c18bac)

key_name	172.24.4.5， 172.24.3.9
metadata	-
name	{}
os-extended-volumes:volumes_attached	Cloud_test
progress	[]
security_groups	0
status	default
tenant_id	ACTIVE
updated	a3c430debe7e48d4a2f58c96dd89746b
user_id	2016-08-1eT06:35:21Z
	db21d0b623e84e48bb1f92b281dcbfa3

查看实例具体的配置信息，具体命令与执行结果如下所示。

```
# nova flavor-show m1.small
```

Property	Value
OS-FLV-DISABLED: disabled	False
OS-FLV-EXT-DATA:ephemeral	0
disk	50
extra specs	{}
id	c7bf1a9c-fdd3-471f-af10-ca34c9402320
name	m1.small
os-flavor-access: is_public	True
ram	1024
rxtx_factor	1.0
swap	
vcpus	1

打开 PuTTY，连接到 Compute 节点，查看其具体的模版信息。

```
# virsh list -all
Id      Name       State
------------------------------------------------------------
3        instance-00000003            running
4        instance-00000004            running

# virsh edit instance-00000003
```

输出详情见附录九 virsh-list.txt。

项目五 网络服务

OpenStack 网络服务为 OpenStack 环境中的虚拟网络基础架构(Virtual Networking Infrastructure，VNI)和物理网络基础架构(Physical Networking Infrastructure，PNI)管理所有的网络现状。OpenStack 网络允许租户创建高级的虚拟网络拓扑结构，包括 Firewalls、Load Dalancers 和 Virtual Private Networks (VPNs)等服务。在成功创建了实例之后，就要考虑实例之间的通信问题，以及实例之间的资源隔离和管理员的访问问题。下面将继续深入学习 OpenStack 网络服务 Neutron 的管理和使用，了解其基本架构和管理方法。

学习目标

- 了解网络服务的相关概念。
- 了解网络服务的基本架构和管理方法。
- 掌握如何创建网络服务。

任务 Neutron 网络管理

任务要求

在 OpenStack 中配置网络是一个令人困惑的经历。它提供了相当数量的功能和灵活性，包括各种新兴的产品插件来支持虚拟网络。接下来，小李就要了解网络服务的相关理论知识，熟悉 Neutron 网络管理命令，然后按照公司需求，创建企业内部网络，满足企业正常的办公需求。

公司要求如下，创建项目研发部内部使用网络，名称为“RD-Net”，子网名为“RD-Subnet”，网段为 172.24.3.0/24，网关为 172.24.3.1。创建业务部内部使用网络，名称为“BS-Net”，子网名为“BS-Subnet”，网段为 172.24.4.0/24，网关为 172.24.4.1。创建 IT 工程部内部使用网络，名称为 IT-Net，子网名为“IT-Subnet”，网段为 172.24.5.0/24，网关为 172.24.5.1。创建外来人员使用网络名称为“Guest-Net”，子网名为“Guest-Subnet”，网段为 172.24.6.0/24，网关为 172.24.6.1。

相关知识

1. 网络服务概述

OpenStack 网络服务为 OpenStack 环境中的虚拟网络基础架构（VNI）和物理网络基础架构接入层方面（PNI）管理所有的网络现状。OpenStack 网络允许租户创建高级的虚拟网络拓扑结构，包括 Firewalls、Load Dalancers 和 Virtual Private Networks（VPNs）等服务。

网络提供了以下对象：网络、子网和路由器。每一个都有模仿其物理副本的功能，网络包含子网，路由器在不同的子网和网络之间进行路由通信。

任何给定的网络设置至少有一个外部网络，这个网络不像其他的网络，它不仅仅是一个实质定义的网络，相反，它代表了到一部分外部网络的视图，可从 OpenStack 安装之外访问到。外部网络的 IP 地址可被任何外部网络的人访问。因为这个网络仅仅代表外部网络的一部分，这个网络上的 DHCP 是禁用的。

除了外部网络，每个网络设置都有一个或更多的内置网络。这些软件定义网络直接连接到虚拟机上。只有内部网络，或者那些在子网通过接口连接到一个类似的路由器的 VMS，可以直接访问连接到那个网络的虚拟机。

为了使外部网络访问 VMS，反之亦然，网络之间的路由器是必需的。每个路由器都有一个连接到网络和许多连接到子网接口的网关。就像一个物理路由器，子网可以访问连接到同一个路由器上的其他子网上的机器，并且机器可以通过路由器的网关访问外部网络。

另外，用户可以分配外部网络的 IP 地址给内置网络的端口。无论什么时候连接到一个子网，该连接就被称为端口。用户可以把外部网络 IP 地址和端口分配给 VMS。通过这种方式，外部网络的实体可以访问虚拟机。

网络也支持安全组。安全组允许管理员在组内定义防火墙规则。一个 VM 可以属于一个或多个安全组，网络运用这些安全组的规则来阻止或开启端口，设置端口范围或 VM 的流量类型。

网络使用的每个插件都有自己的概念。虽然不是至关重要的操作网络，但是理解这些概念可以帮助用户建立网络。所有的网络设备都要使用一个核心插件和一个安全组插件（或只是空操作安全组插件）。此外，firewall-as-a-service（FWaaS）和 load-balancing-as-a-service（LBaaS）插件可用。

Neutron 网络服务为 OpenStack 管理所有的网络方面的虚拟网络基础架构和访问层方面的物理网络基础架构。在 OpenStack 中，计算、存储和网络是其核心内容，也是核心组件，可通过具体的功能实现和服务访问，提供云计算环境下的虚拟网络功能，如图 5-1 所示。Neutron 为 OpenStack 的虚拟机提供网络方面的功能，在 Havana 版本之前是没有 Neutron 这个组件（G 版名称是 Quantum）的，当时网络的主要功能是在 Nova 组件里实现的，即 Nova Network，底层采用的大多是 Linux Bridge，但是无法快速组网和实现高级的网络功能，因此 OpenStack 把 Nova 的关于网络方面的功能进行转移，成立了全新的 Neutron 组件，Neutron 组件取代了 Nova Network 的相关功能，但是 Nova 里还有些网络功能被保留了，如虚拟机的网卡方面的功能。在 OpenStack 中，网络功能是最复杂的功能，很多计算和存储方面的问题都是和网络紧密相关的。

目前，对于我们使用的物理交换机来说，Neutron 其实是系统平台的位置，提供配置命令及参数检查，并把网络功能用一种逻辑组织起来，但是，无论是用底层的 Plugin 还是最终用软件 SDN 通过硬件交换机来加速，Neutron 自身并不提供任何网络功能，它只是一个架子。Neutron 的网络功能大部分是 Plugin 提供的。

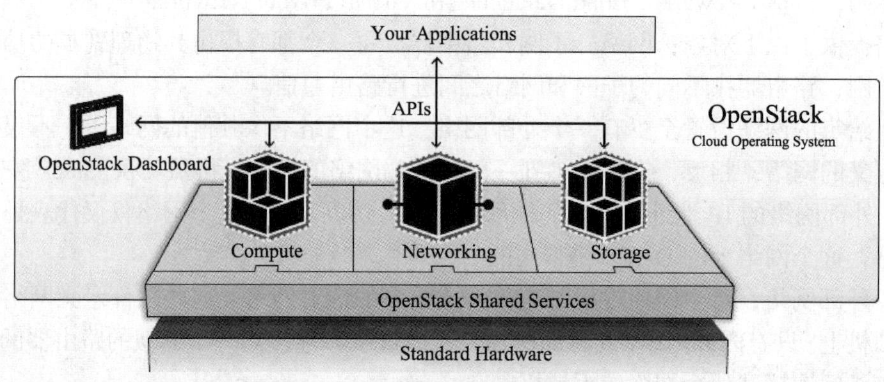

图 5-1　OpenStack 中的 Neutron

2. 网络服务架构介绍

Neutron 网络的目的是灵活地划分物理网络，在多租户环境下提供给每个租户独立的网络环境。它是可以被用户创建的对象，如果要和物理环境下的概念映射的话，这个对象相当于一个巨大的交换机，可以拥有无限多个动态可创建和销毁的虚拟端口。

从本质上讲，OpenStack 网络创建了一个一致的逻辑通信层，而其他元素可以有效地、大规模地使用这个逻辑层，这一模式是在 OpenStack Neutron 服务器中实现的，它与 OpenStack Nova 软件管理的虚拟机（即计算服务），以及上述提及的其他元素进行交互（Neutron 是由 Nova Network 演变而来的）。用户通过 OpenStack 的 Horizon GUI 与网络功能进行交互，而其他的管理系统和网络则通过使用 Neutron API 与网络服务进行交互。在目前的分布中，OpenStack Neutron 混合实施了第 2 层的 Vlan 和第 3 层的路由服务，它可为所支持的网络提供防火墙、负载平衡以及 IPSec VPN 等扩展功能。

网络中的虚拟机来自于虚拟机管理程序，其中包括 KVM、ESX、XenServer 和 Hyper-V，它支持设计中的可选择性、多功能性。通过使用支持 IPv4 和 IPv6 的强大地址管理功能，Neutron 可在一个站点内实现虚拟机的灵活部署。它采用了一个逻辑的子网架构，IP 块表示可以寻址虚拟位置和到达其部署结构中任何的位置。这种结构类似于在众多 Neutron IP 组网中所使用的子网技术，它通常是从与子网相关的物理端口抽取出来的。在覆盖网络中，它可被用作逻辑关联机制，以便为不同用户和应用程序分离网络流量。通过使用这种方法，云计算供应商可以分别实现多个租户的通信并为不同应用程序部署拓扑结构，例如，多层服务器组合以处理一个整体应用程序的不同部分，目前 Neutron 网络形式主要包括 Flat、Vlan、Gre、VXlan。

对于 Neutron 来说，它相当于 OpenStack 的网络管理平台，其可以提供 CLI 或者 GUI 端的配置或者对外提供 API 访问接口，把底层的网络统一组织起来。Neutron 本身并不对外提供任何网络服务，其提供一个底层网络设备和一个访问接口的框架。Neutron 组件可以分为 5 个主要子服务，分别为 Neutron Server、OpenvSwitch Agent、DHCP Agent、L3 Agent 和 Metadata Agent，如图 5-2 所示。

（1）Neutron Server

实现 Neutron API 和 API 扩展，管理 Network、Subnet、Port。

（2）OpenvSwitch Agent

连接虚拟机到网络端口，运行在每个计算节点。

（3）DHCP Agent

负责 DHCP 配置，为虚拟机分配 IP。

（4）L3 Agent

负责公网浮动 IP 地址和 NAT，负责其他 3 层特性，如负载均衡，每个 Network 对应一个 L3 Agent。

Metadata Agent
Neutron Sever
Queue
DB
L2 Agent
L3 Agent
OpenvSwitch Agent
DHCP Agent

图 5-2　Neutron 架构示意图

（5）Metadata Agent

提供元数据服务，如 Neutron 的 L3 Agent、DHCP Agent、Nova、Metadata API Server。

在默认情况下，Neutron 使用 OVS（OpenvSwitch）作为其虚拟交换的 Plugin 插件，同时租户子网的创建和租户之间以及租户与外部之间的通信均由 Neutron 创建的网络实现，如图 5-3 所示，由 OVS 创建 3 个网段，租户的实例资源连接到对应的租户网段，从而实现租户网段的网络畅通。

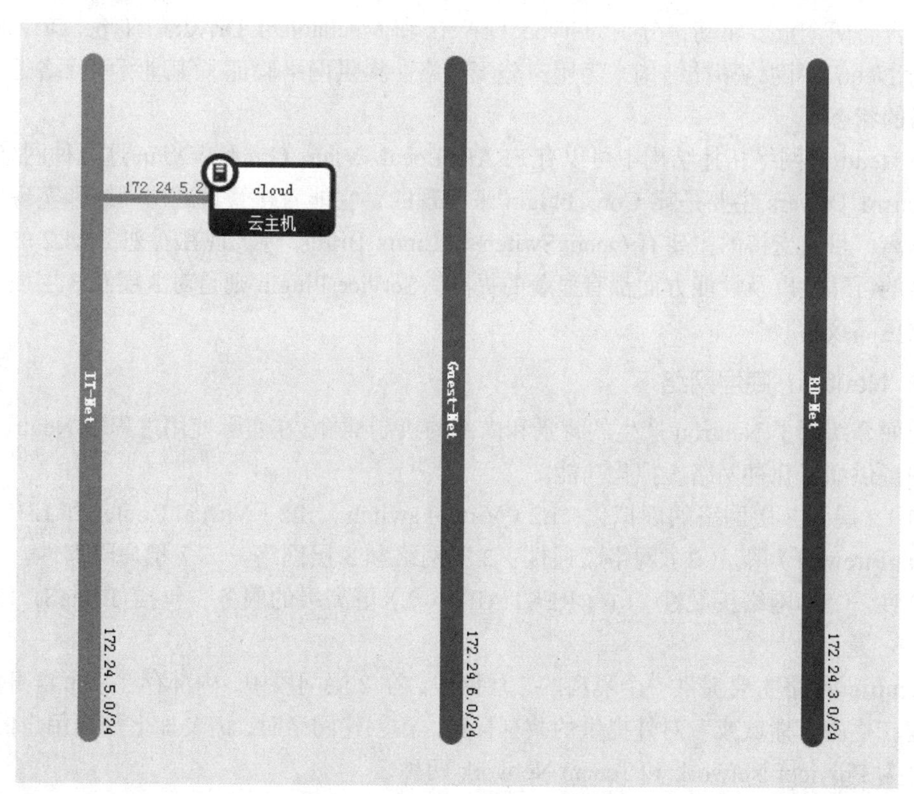

图 5-3　Neutorn 网络拓扑结构

一个标准的 Neutron 网络是由 Network、Subnet、Router 和 Port 组成的。

（1）Network

2 层网络单元，租户可以通过 Neutron 的 API 创建自己的网络，类似 Vlan 的 3 层接口，接口则对应 2 层的一个单独的隔离网络。

（2）Subnet

一段 IPv4/IPv6 地址段，提供公网或者私网的地址，相当于在 Network 的基础上，2 层的单独隔离网络，VM 同属于一个 IP 段，具有同样的网络掩码，具备相同的网络网关。

（3）Router

虚拟 3 层路由器，为私网地址提供地址转换功能。

（4）Port

管理实例的网卡，虚拟机交换机的一个接口。

在 Neutron 配置文件中同时有两种不同的 Plugin，分别为 Core Plugin 和 Service Plugin，其中，Core Plugin 为实现 2 层网络中网络之间的隔离和互通，默认配置为 ML2（Modular Layer 2）；Service Plugin 则提供 3 层及 3 层以上的服务，如 Router、FWaas、LBaas 和 VPNaas。

从 Core Plugin 角度来说，目前的操作只是操作子网、管理子网和管理实例连接 OVS 的虚拟接口，而在实际情况下它主要和 Neutron 的 Agent 通信，Agent 则调用底层所支持的 Plugin 去执行终端用户的命令，在 OpenStack 中常用的 ML2 插件就是一个标准的 Core Plugin 插件，它用于在 Linux Bridge 和 OpenvSwitch 之间实现具体的网络管理；在 ML2 中，它的内部也含有两个重要驱动，分别为 Type_Drivers 和 Mechanism_Drivers，Type_Drivers 主要是根据 Neutron 网络拓扑结构来为用户分配网络、提供网络验证、管理可用网络类型和维护网络的状态。

在 Neutron 网络拓扑结构中可以有 FLAT、Local、Vlan、Gre 和 VXlan 这几种网络模式。Mechanism_Drivers 主要是靠 Core Plugin 来实现的，它能直接管理底层网络和实现底层网络的隔离，目前支持的主要有 OpenvSwitch 和 Linux Bridge 等。现在看来，ML2 的出现使得网络的可靠性以及性能方面都有显著的提高。Service Plugin 则是对 3 层及 3 层以上的网络提供更高级的服务。

3. Neutron 底层网络

前面介绍到了 Neutron 产生的背景和内部的机制插件，在实际使用过程中 Neutron 主要向用户提供如下几种网络虚拟化功能。

（1）2 层到 7 层网络的虚拟化：L2（Virtual switch）、L3（Virtual Router 和 LB）、L4-7（Virtual Firewall）等。（2）网络联通性：2 层网络和 3 层网络。（3）租户隔离性。（4）网络安全性。（5）网络扩展性。（6）REST API。（7）更高级的服务，包括 LBaaS，FWaaS，VPNaaS 等。

Neutron 网络主要提供一个隔离的 2 层网段，在 2 层网段中，内部存在一个自身内部租户创建产生的广播域或者对外提供的共享标记。在创建网络时，需要加上创建用户的权限，可以分为 Provider Network 和 Tenant Network 两种。

在 Provider Network 中，网络均由管理员创建，对应于数据中心已有的物理网络中的一个网段，常见的创建为 provider:network_type（网络类型，包括 VXlan、Gre、Vlan、Flat、Local），当然这种网络仅仅只是作为租户或者租户之间共享的网络，并不需要 L3 Agent。

在 Tenant Network 中，网络均是由租户的普通用户创建的，默认情况下，租户是不能创建共享的网络的，所以要保证租户之间的网络隔离，不被其他租户发现和使用。在 Tenant

Network 中也有 Local、Flat、Vlan、Gre 和 VXlan 等类型。但是，Tenant 普通用户创建的 Flat 和 Vlan 的 Tenant Network 实际还是 Provider Network（属于某一个实际物理网段），所以真正产生 Tenant Network 的为 Gre 和 VXlan，这种和物理网络没有任何的联系。

Provider Network 与 Tenant Network 有以下几点区别。

① Provider Network 是由“admin”用户创建的，而 Tenant Network 是由 Tenant 普通用户创建的。

② Provider Network 和物理网络的某段直接映射，例如，对应某个 Vlan，需要预先在物理网络中做相应的配置。而 Tenant Network 是虚拟化的网络，Neutron 需要负责其路由等 3 层功能。

③ 对 Flat 和 Vlan 类型的网络来说，只有 Provider Network 才有意义。即使是这种类型的 Tenant Network，其本质上也是对应于一个实际的物理段。

④ 对 Gre 和 VXlan 类型的网络来说，只有 Tenant Network 才有意义，因为它本身不依赖于具体的物理网络，只需要物理网络提供 IP 和组播即可。

⑤ Provider Network 根据“admin”用户输入的物理网络参数创建，而 Tenant Network 由 Tenant 普通用户创建，Neutron 根据其网络配置来选择具体的配置，包括网络类型、物理网络和 segmentation_id。

⑥ 创建 Provider Network 时允许使用不在配置项范围内的 segmentation_id，如图 5-4 所示。

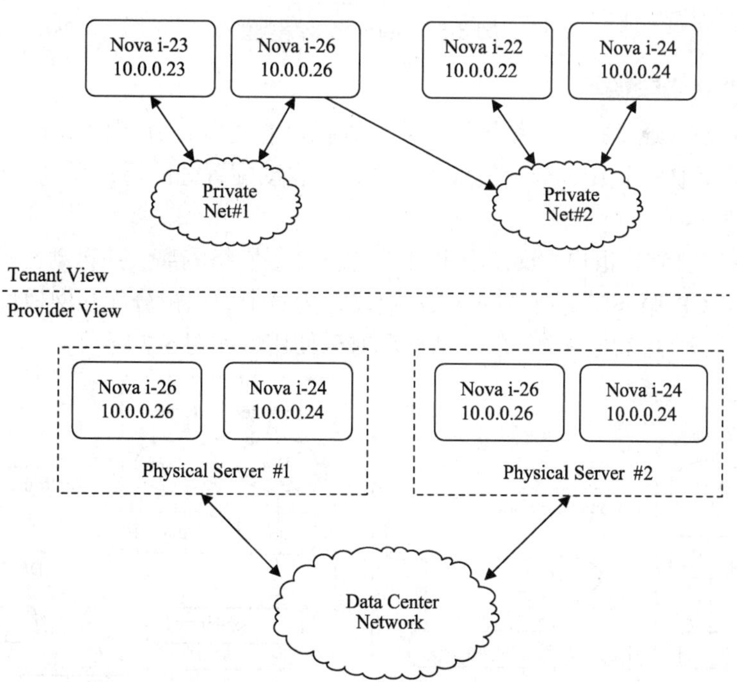

图 5-4　Neutron 创建 Provider

在一个实际的生产环境中，Neutron 的网络可以分为以下几种：实例通信网络、内部存储网络、内部管理网络和实例私有网络。

实例通信网络为外部通信网络；内部管理网络即 OpenStack 自身组件所使用的网络，

通常为管理组件之间通信所使用的网络；实例私有网络为虚拟机使用的内部网络；内部存储网络为存储服务器之间的通信网络，负责数据之间的数据交换和数据备份，如图 5-5 所示。

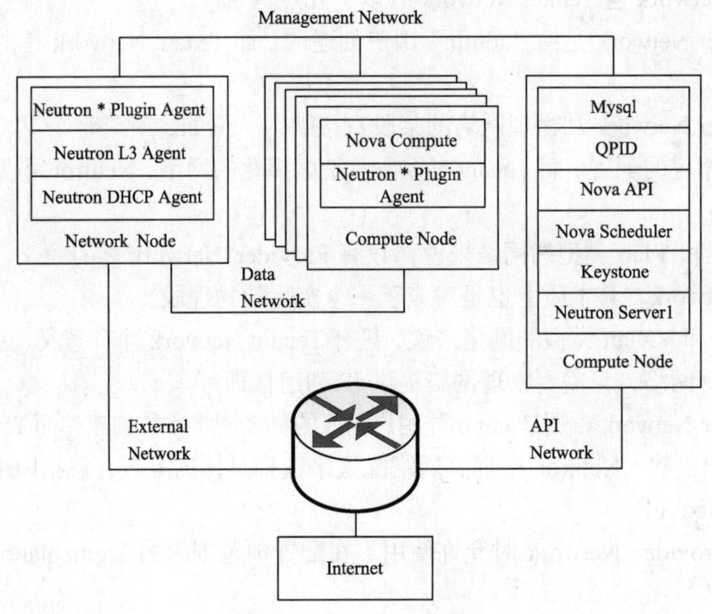

图 5-5 Neutron 网络类型

4. Neutron 网络模式

FlatDHCP 模式通常指定一个子网，规定虚拟机能使用的 IP 地址范围，创建实例时从有效 IP 地址池获取一个 IP，分配给虚拟机实例，自动配置好网桥，通过 DNSmasq 分配地址，如图 5-6 所示。

Vlan 模式需要创建租户 Vlan，使得租户之间 2 层网络隔离，并自动创建网桥，在网络控制器上的 DHCP 服务器为所有的 Vlan 启动，为每个虚拟机分配私网地址。网络控制器 NAT 转换，也可解决 2 层隔离问题，适合私有云使用，如图 5-7 所示。

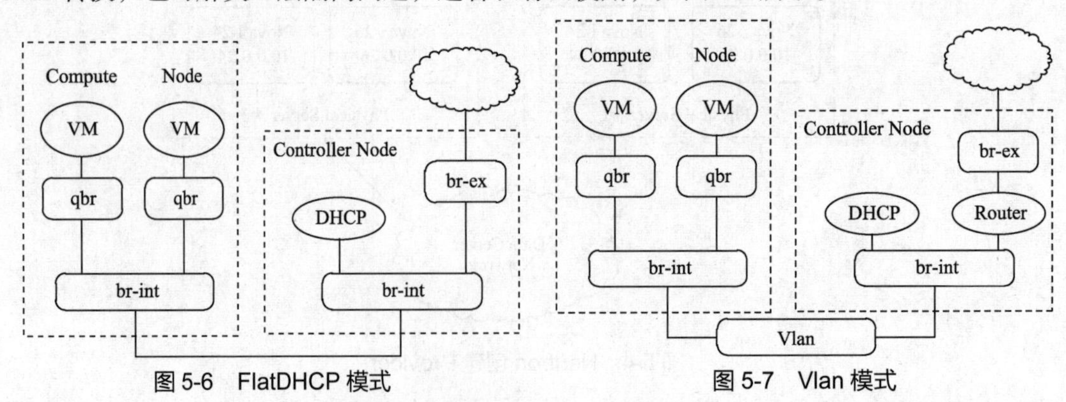

图 5-6 FlatDHCP 模式　　图 5-7 Vlan 模式

Gre 网络可以跨不同网络实现二次 IP 通信，而且通信封装在 IP 报文中，可实现点对点隧道，如图 5-8 所示。

5. 数据包接收

在 Linux 网络系统中，部分网卡自带端口自协商的功能，设备双方通过这样的方式建立合适的数据连接，包括连接速度和连接方式。在数据接收方面，将数据内容放置在程序缓存部分，而后通过中断方式通知 CPU/CPU 轮询读取数据帧；在数据发送方面，是以软中断方式传递给发包队列，而后匹配驱动才能让 CPU 控制网卡设备将数据发送出去。

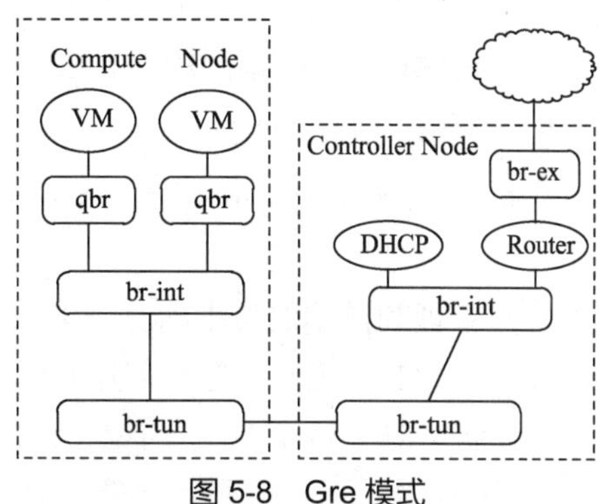

图 5-8 Gre 模式

通过网卡的设置不同和报文的目的 MAC 地址可判断是否接收该报文。

网卡模式主要以下几种。

（1）直接模式

只接收目的 MAC 地址是网卡自身地址的数据帧。

（2）组播模式

接收所有的组播报文数据帧。

（3）广播模式

接收广播帧。

（4）混杂模式

可以接收所有任意目的 MAC 地址的报文，无论报文的目的 MAC 地址是否是自身的网卡 MAC 地址，无论是单播或者非单播。

通常情况下网卡只接收 MAC 地址为网卡自身地址的数据帧和广播的数据帧，以及自己所在组播的组播报文。

当网卡被加入到 OpenvSwitch 中作为一个端口时会被设置成混杂模式。

6. Linux Bridge 和 Vlan

Linux Bridge 模拟了物理网络中的网桥的概念，将多个服务器端口加入到网桥中来，网桥端口对端相连的设备通过发送报文给 Linux Bridge，并通过 Linux Bridge 学习报文的 SMAC 和查找报文的 DMAC 转发到相应的目的地，这点也和普通的 2 层交换机非常类似。Linux Bridge 既可以是物理网卡设备也可以是虚拟的端口。

Linux Bridge 本身是没有 Vlan 功能的，需要 Vlan 模块协作实现 Vlan 过滤的功能，当端口加入到 Linux Bridge 中时，默认是不属于任何 Vlan 的，只有这样才能对所有的 Vlan 的报文进行接收和转发，这和普通的交换机的 Trunk 端口还是有区别的。交换机的 Trunk 是针对 Vlan 的端口属性，需要自己配置 Vlan 才能使转发报文通过；Linux Bridge 如果没有配置 Vlan 端口就会让所有的报文通过。

Linux Bridge 在 OpenStack 的云计算网络中也有着非常重要和广泛的应用。Nova Network 就用 Linux Bridge 和 Vlan 做租户隔离，即使在后面的 Neutron 组建中，OpenvSwitch 做了 Plugin 时也大量用到了 Linux Bridge 和 Vlan。

7. OpenvSwitch 说明

OpenvSwitch 是开源 Apache 2.0 的高质量多层次虚拟交换机软件，在某一个虚拟化的环境中，Open vSwitch 的主要功能是传递 VM 之间的流量，实现 VM 和外界网络通信。最初的虚拟机之间的隔离使用了 Nova Network，底层采用的是 Linux Bridge 技术，虚拟机网络的组建和配置非常不灵活而且配置无法模块化，扩展非常困难，对于运维来说对错误定位也有一定的难度，OpenvSwitch 就是为了解决以上问题而出现的，它支持 QoS、镜像和 NetFlow 功能等。OpenvSwitch 支持 Vlan、VXlan 和 Gre 等多种隔离方式，同时支持 OpenFlow 协议，在后续的 SDN 技术中可以延续对其的管理。

8. NameSpace 方案

Linux 系统中的 NameSpace 称之为 container，是一种资源隔离方案，是基于容器的虚拟化技术的基础，和 C++里面的 NameSpace 非常类似，所以 NameSpace 是属于 Linux 系统层次资源管理和使用的技术内容，不属于网络部分的概念。

在 Linux 的系统调用进程创建函数 clone，设置一个 CLONE_NEWPID，参数就会新建一个 NameSpace，每个 NameSpace 就如同一个新系统，进程号也是从 1 开始，它拥有独立的系统 PID、IPC 通信机制、网络环境、路由表和防火墙规则。

（1）NameSpace 特点

① 在不同的 NameSpace 中资源是隔离的，信息传递只能通过网络来完成。

② 部署 NameSpace 开销小，部署迅速，但是这对内核开发来说是难事。

③ 与其他的虚拟化技术和云计算平台相比，不能指定任意想要的操作系统并部署想要的操作软件。

（2）Linux NameSpace 的常见命令

① 显示信息：包括路由和 DHCP 信息。执行结果如下。

```
# ip netns list
qdhcp-d152d61e-a548-492e-a634-23c87b91f1eb
qrouter-5bbd40a9-e02c-440d-96e3-f7b40ba50cce
qdhcp-1994aba9-ea74-4a8c-985b-c74cf6c8cb34
qdhcp-97c9c430-a5a0-490f-9d2f-dbf8d796ff2a
```

② 显示某一个 NameSpace 内部的信息，如显示网络地址，命令及结果如下。

```
# ip netns exec qdhcp-97c9c430-a5a0-490f-9d2f-dbf8d796ff2a ip a
14: lo: <LOOPBACK,UP,LOWER_UP>mtu 16436 qdiscnoqueue state UNKNOWN
link/loopback 00:00:00:00:00:00 brd 00:00:00:00:00:00
inet 127.0.0.1/8 scope host lo
inet6 ::1/128 scope host
valid_lft forever preferred_lft forever
```

③ 查看防火墙信息，命令及结果如下。

```
# ip netns exec qrouter-5bbd40a9-e02c-440d-96e3-f7b40ba50cce iptables -nL
Chain INPUT (policy ACCEPT)
targetprot opt source             destination
neutron-l3-agent-INPUT  all  --  0.0.0.0/0          0.0.0.0/0
```

```
Chain FORWARD (policy ACCEPT)
targetprot opt source              destination
neutron-filter-top  all  --  0.0.0.0/0            0.0.0.0/0
neutron-l3-agent-FORWARD  all  --  0.0.0.0/0            0.0.0.0/0

Chain OUTPUT (policy ACCEPT)
targetprot opt source              destination
neutron-filter-top  all  --  0.0.0.0/0            0.0.0.0/0
neutron-l3-agent-OUTPUT  all  --  0.0.0.0/0            0.0.0.0/0

Chain neutron-filter-top (2 references)
targetprot opt source              destination
neutron-l3-agent-local  all  --  0.0.0.0/0            0.0.0.0/0

Chain neutron-l3-agent-FORWARD (1 references)
targetprot opt source              destination

Chain neutron-l3-agent-INPUT (1 references)
targetprot opt source              destination
ACCEPT     tcp  --  0.0.0.0/0            127.0.0.1            tcp dpt:9697

Chain neutron-l3-agent-OUTPUT (1 references)
targetprot opt source              destination

Chain neutron-l3-agent-local (1 references)
targetprot opt source              destination
```

④ 创建 NameSpace 测试网桥：模拟 Linux 网桥的创建，并结合 OpenvSwitch 的网桥，完成一个测试网桥，并连接两个 NameSpace。预期完成的效果如图 5-9 所示。

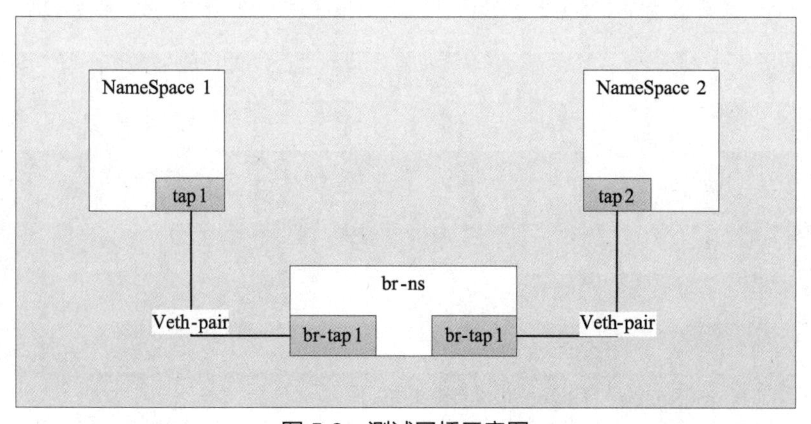

图 5-9　测试网桥示意图

- 创建 namespace1 和 namespace2 命名空间。

```
# ip netns add namespace1
# ip netns add namespace2
```

- 创建 br-ns 网桥。

```
# ovs-vsctl add-brbr-ns
```

- 创建内部通信端口。

```
# ovs-vsctl add-port br-ns tap1 -- set interface tap1 type=internal
# ovs-vsctl add-port br-ns tap2 -- set interface tap2 type=internal
```

- 将 tap1 和 tap2 分别加入命名空间。

```
# ip link set tap1 netns namespace1
# ip link set tap2 netns namespace2
```

- 配置命名空间地址。

```
# ip netns exec namespace1 ipaddr add 192.168.0.10/24 dev tap1
# ip netns exec namespace1 ip link set tap1 up
# ip netns exec namespace1 ip link set lo up
# ip netns exec namespace2 ipaddr add 192.168.0.20/24 dev tap2
# ip netns exec namespace2 ip link set tap2 up
# ip netns exec namespace2 ip link set lo up
```

- 检测网络是否通畅。

```
# ip netns exec namespace1 ping 192.168.0.10
# ip netns exec namespace2 ping 192.168.0.20
# ip netns exec namespace1 ping 192.168.0.20
# ip netns exec namespace2 ping 192.168.0.10
```

- 创建 OVS，输出并查看，效果如下。

```
[root@controller ~]# ovs-vsctl show
a624a3b6-63ab-4ffd-b7ca-503277ec222d
   Bridge br-ns
       Port "tap2"
           Interface "tap2"
type: internal
       Port "tap1"
           Interface "tap1"
type: internal
       Port br-ns
           Interface br-ns
type: internal
ovs_version: "2.1.3"
```

- 查看“NameSpace”命令，执行效果如下。

```
[root@controller ~]# ip netns list
```

```
namespace1
namespace2
```

- 设定查看某个 NameSpace 的地址，执行效果如下所示。

```
[root@controller ~]# ip netns exec namespace1 ip a
28: lo: <LOOPBACK,UP,LOWER_UP>mtu 16436 qdiscnoqueue state UNKNOWN
link/loopback 00:00:00:00:00:00 brd 00:00:00:00:00:00
inet 127.0.0.1/8 scope host lo
inet6 ::1/128 scope host
valid_lft forever preferred_lft forever
31: tap1: <BROADCAST,UP,LOWER_UP>mtu 1500 qdiscnoqueue state UNKNOWN
link/ether f6:88:67:c6:85:99 brdff:ff:ff:ff:ff:ff
inet 192.168.0.10/24 scope global tap1
   inet6 fe80::f488:67ff:fec6:8599/64 scope link
valid_lft forever preferred_lft forever
```

9. DNSmasq 工具

DNSmasq 是一个小巧且方便，用于配置 DNS 和 DHCP 的工具，适用于小型网络，它提供了 DNS 功能和可选择的 DHCP 功能。它服务那些只在本地适用的域名，这些域名是不会在全球的 DNS 服务器中出现的。DHCP 服务器和 DNS 服务器结合，并且允许 DHCP 分配的地址能在 DNS 中正常解析，而这些 DHCP 分配的地址和相关命令可以配置到每台主机中，也可以配置到一台核心设备中(比如路由器)，DNSmasq 支持静态和动态两种 DHCP 配置方式。

查看已经启动的 DNSmasq 进程的命令如下。

```
# ps aux |grep dnsmasq
nobody    1218  0.0  0.0  12884   796 ?        S    15:10   0:00 dnsmasq
--no-hosts --no-resolv --strict-order --bind-interfaces --interface= ap689d790c-
72  --except-interface=lo  --pid-file=/var/lib/neutron/dhcp/d152d61e-a548-492e-
a634-23c87b91f1eb/pid --dhcp-hostsfile=/var/lib/neutron/dhcp/d152d61e -a548-492e-
a634-23c87b91f1eb/host   --addn-hosts=/var/lib/neutron/dhcp/d152d61e-a548-492e-
a634-23c87b91f1eb/addn_hosts  --dhcp-optsfile=/var/lib/neutron/dhcp/d152d61e-
a548-492e-a634-23c87b91f1eb/opts  --leasefile-ro  --dhcp-range=tag0,10.1.1.0,
static,86400s --dhcp-lease-max=16 --conf-file=/etc/neutron/dnsmasq- neutron.
conf --domain=openstacklocal
nobody    2230  0.0  0.0  12888   696 ?        S    Oct11   0:00 /usr/
sbin/dnsmasq --strict-order --local=// --domain-needed --pid-file=/ var/ run/
libvirt/network/default.pid  --conf-file=  --except-interface  lo  --bind-
interfaces  --listen-address 192.168.122.1  --dhcp-range 192.168.122.2,192.
168.122.254 --dhcp-leasefile=/var/lib/libvirt/dnsmasq/default.leases --dhcp-
lease-max=253  --dhcp-no-override  --dhcp-hostsfile=/var/lib/libvirt/dnsmasq/
```

```
default.hostsfile --addn-hosts=/var/lib/libvirt/dnsmasq/default.addnhosts
nobody    2783  0.0  0.0  12884   728 ?        S    Oct11   0:00 dnsmasq
--no-hosts --no-resolv --strict-order --bind-interfaces --interface=tap42098fe2-
05 --except-interface=lo --pid-file=/var/lib/neutron/dhcp/1994aba9-ea74-4a8c-
985b-c74cf6c8cb34/pid --dhcp-hostsfile=/var/lib/neutron/dhcp/1994aba9-ea74-4a8c-
985b-c74cf6c8cb34/host --addn-hosts=/var/lib/neutron/dhcp/1994aba9- ea74-4a8c-
985b-c74cf6c8cb34/addn_hosts --dhcp-optsfile=/var/lib/neutron/dhcp/ 1994aba9-
ea74-4a8c-985b-c74cf6c8cb34/opts --leasefile-ro --dhcp-range=tag0, 51.0.0.0,
static,86400s --dhcp-lease-max=256--conf-file=/etc/neutron/dnsmasq- neutron.conf
-- domain=openstacklocal
nobody    2785  0.0  0.0  12884   732 ?        S    Oct11   0:00 dnsmasq
--no-hosts --no-resolv --strict-order --bind-interfaces --interface=tapfa7371e3-
b4 --except-interface=lo --pid-file=/var/lib/neutron/dhcp/97c9c430-a5a0-490f-
9d2f-dbf8d796ff2a/pid --dhcp-hostsfile=/var/lib/neutron/dhcp/97c9c430-a5a0-
490f-9d2f-dbf8d796ff2a/host --addn-hosts=/var/lib/neutron/dhcp/97c9c430-a5a0-
490f-9d2f-dbf8d796ff2a/addn_hosts --dhcp-optsfile=/var/lib/neutron/dhcp/97c9c430-
a5a0-490f-9d2f-dbf8d796ff2a/opts --leasefile-ro --dhcp-range=tag0,50.0.0.0,
static,86400s --dhcp-lease-max=256 --conf-file=/etc/neutron/dnsmasq-neutron.conf
-- domain=openstacklocal
root     6696  0.0  0.0 103248   872 pts/2    S+   16:00   0:00 grepdnsmasq
```

通过上面的结果可以看出，它的主要作用是为虚拟机分配 IP 地址，即每一个 DNSmasq 进程提供一个 DHCP 服务，同样地，每启动一个这样的 DHCP 进程，都保存在网络节点的/var/lib/neutron/dhcp/目录内部，而实例对应的 MAC 地址和分配的 IP 地址则对应在网络 ID 内的 hosts 文件内。

```
[root@controller ~]# ls -l /var/lib/neutron/dhcp/
total 4
drwxr-xr-x. 2 neutron neutron 4096 Oct 12 15:30 d152d61e-a548-492e-a634-
23c87b91f1eb
[root@controller  ~]#  cat  /var/lib/neutron/dhcp/d152d61e-a548-492e-a634-
23c87b91f1eb/host
fa:16:3e:1b:66:b0,host-10-1-1-2.openstacklocal,10.1.1.2
fa:16:3e:fc:59:f3,host-10-1-1-1.openstacklocal,10.1.1.1
fa:16:3e:8a:9c:3b,host-10-1-1-4.openstacklocal,10.1.1.4
```

10. Neutron 网络拓扑结构

在多平面网络中，存在实例网络和外部通信网络同为一个网络，但同时又可以共享多个网络的情况，当两个虚拟机需要相互通信时，虚拟机内部可以通过多个网卡同时共享一个物理网段，如图 5-10 所示。

混合平面私有网络是指每个 Tenant 拥有租户自己的实例网络，同时共享一个相同的共享实例外部网络。其中，第一台VM同属于Tenant A和Tenant B，第二台VM则只属于Tenant B私有网络，如图5-11所示。

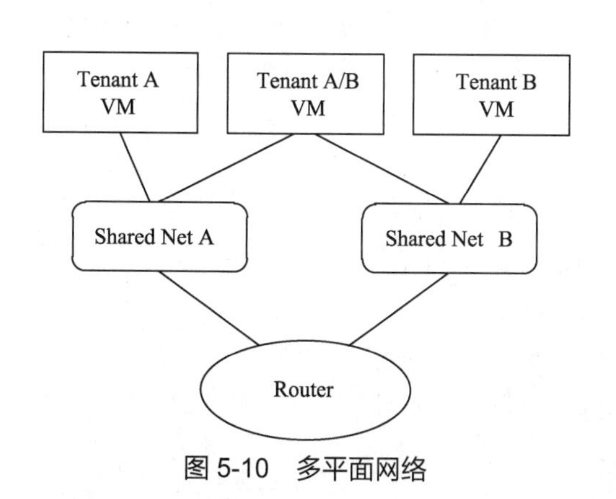

图5-10 多平面网络

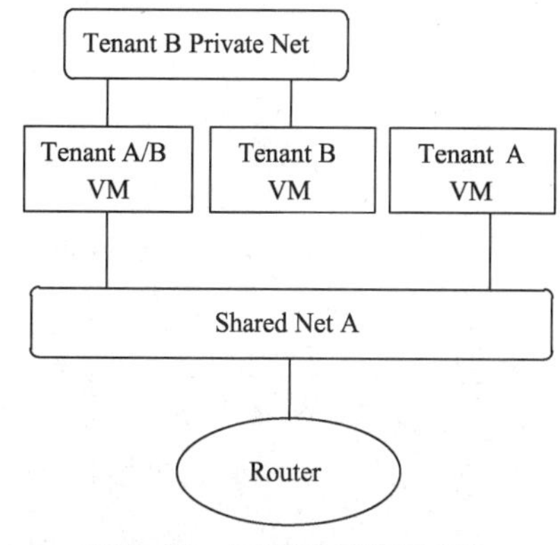

图5-11 混合平面私有网络

私有网络可实现运营商路由功能。通过虚拟路由功能，可连接多个租户的内部实例网络，同时可以直接通过路由提供的SNAT功能访问外部网络，如图5-12所示。

通过私有网络，每个租户可创建自己专属的网络区段，每个租户对应租户路由设备，创建自己的专属网络环境，每个网络连接自己专属的虚拟路由器，满足访问外部网络的需求，如图5-13所示。

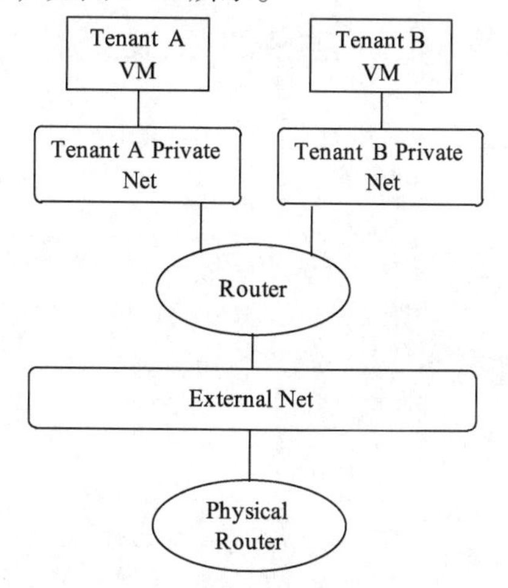

图5-12 私有网络实现运营商路由

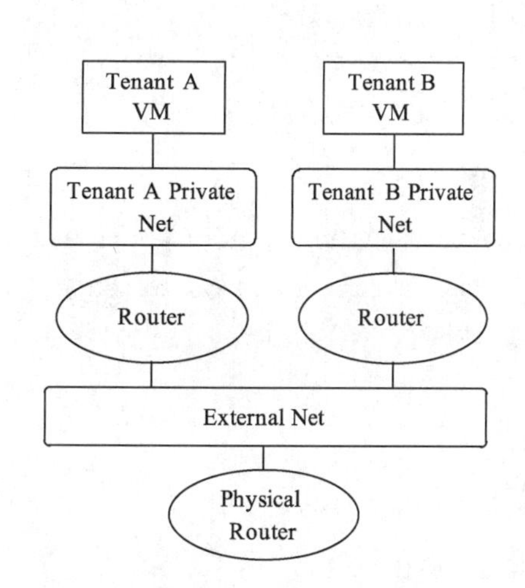

图5-13 私有网络实现每个租户创建自己专属的网络区段

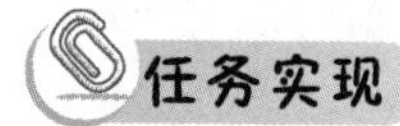

任务实现

1. 基础操作练习

（1）创建网络

列出系统扩展名的命令如下，执行结果如下所示。

```
$ neutron ext-list -c alias -c name
```

生效环境变量的命令如下。

```
# source /etc/keystone/admin-openrc.sh
```

alias	name
security-group	security-group
l3_agent_scheduler	L3 Agent Scheduler
ext-gw-mode	Neutron L3 Configurable external gateway mode
binding	Port Binding
provider	Provider Network
agent	agent
quotas	Quota management support
dhcp_agent_scheduler	DHCP Agent Scheduler
multi-provider	Multi Provider Network
external-net	Neutron external network
router	Neutron L3 Router
allowed-address-pairs	Allowed Address Pairs
extra_dhcp_opt	Neutron Extra DHCP opts
extraroute	Neutron Extra Route

创建网络的命令如下，执行结果如下。

```
$ neutron net-create ext-net --shared --router:external=True
```

Field	Value
admin_state_up	True
id	46ead639-8ae5-401f-af11-370857f09a0b
name	ext-net
provider:network_typeprovider:physical_network	vlan
provider:segmentation_idrouter:external	physnet1
shared	45
status	True
subnets	True
tenant_id	ACTIVE
	a3c430debe7e48d4a2f58c96dd89746b

（2）创建子网

创建子网的命令如下，执行结果如下。

```
$ neutron subnet-create ext-net --name ext-subnet  --allocation-pool start=
172.24.7.100,end=172.24.7.200     --disable-dhcp  --gateway  172.24.7.254
172.24.7.0/24
Created a new subnet:
```

Field	Value
allocation_pools	{"start": "172.24.7.100", "end": "172.24.7.200"}

```
cidr                    | 172.24.7.0/24
dns_nameservers         |
enable_dhcp             | False
gateway_ip              | 172.24.7.254
host_routes             |
id                      | 322d7f3a-c254-4547-9cd1-306ad3bb6b4a
ip_version              | 4
name                    | subnet
network_ip              | 8bbff1f6-66a0-4ffb-87f8-cba4beec27e0
tenant_id               | a3c430debe7e48d4a2f58c96dd89746b
```

（3）创建用户网络

创建用户网络的命令如下，执行结果如下。

```
$ neutron net-create  demo-net
Created a new network:
 Field                          | Value
 admin_state_up                 | True
 id                             | 2b75df30-990f-4ff9-8d12-6fbeb23a0b10
 name                           | demo-net
 provider:network_type          | vlan
 provider:physical_network      | physnet1
 provider:segmentation_id       | 45
 shared                         | False
 status                         | ACTIVE
 subnets                        |
 tenant_id                      | a3c430debe7e48d4a2f58c96dd89746b
```

（4）创建用户子网

创建用户子网的命令如下，执行结果如下。

```
$ neutron subnet-create demo-net --name demo-subnet   --gateway 10.0.0.1
10.0.0.0/24
Created a new subnet:
 Field                  | Value
 allocation_pools       | {"start": "10.0.0.2", "end": "10.0.0.254"}
 cidr                   | 10.0.0.0/24
 dns_nameservers        |
 enable_dhcp            | True
 gateway_ip             | 10.0.0.1
 host_routes            |
 id                     | bb6cc79d-6ef9-4bc8-aa1d-f640dea00594
 ip_version             | 4
 name                   | demo-subnet
```

network_id	2b75df30-990f-4ff9-8d12-6fbeb23a0b10
tenant_id	a3c430debe7e48d4a2f58c96dd89746b

（5）创建路由器

创建路由器的命令如下，执行结果如下。

```
$ neutron router-create router1
Created a new router:
```

Field	Value
admin_state_up	True
external_gateway_info	
id	967a6805-c403-405b-af85-9831c01d9b22
name	router1
status	ACTIVE
tenant_id	a3c430debe7e48d4a2f58c96dd89746b

（6）创建路由网关接口

创建路由网关接口的命令如下，执行结果如下。

```
$ neutron router-interface-add router1 demo-subnet
Added interface 70c3c2af-cc96-4e5c-80aa-d9797656e836 to router router1.
$ neutron router-gateway-set router1 ext-net
Set gateway for router router1
```

（7）查看整体效果

结果如图 5-14 所示。

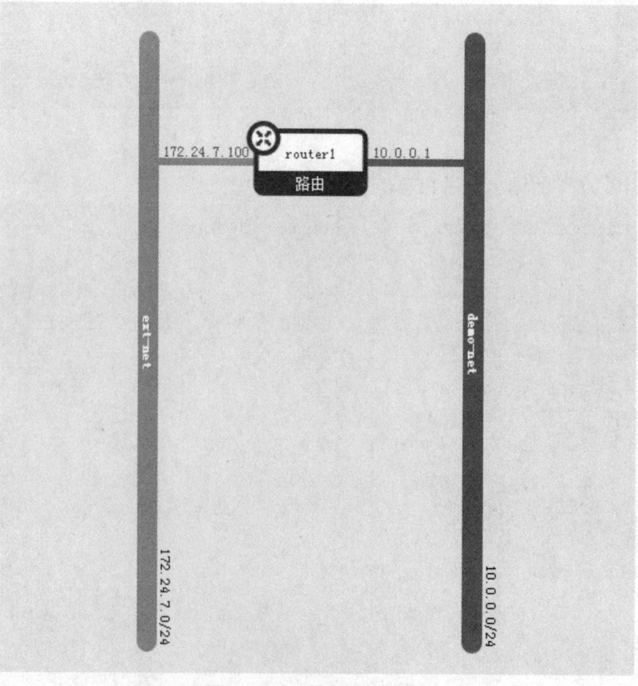

图 5-14　整体效果图

2. 创建各部门网络子网和外来访问使用网络

（1）通过 GUI 界面为项目研发部创建网络和子网

① 进入 Dashboard 界面找到“项目”选项。

② 打开网络面板，找到“网络”。

③ 选择“创建网络”。

④ 在弹出的窗口中的网络标签中输入网络名称，选择是否处于管理员状态，完成后的结果如图 5-15 所示。

⑤ 在子网标签中，输入子网名称、网络地址、IP 版本和网关 IP 等信息。

⑥ 在子网详情标签中，输入分配的地址池，完成后结果如图 5-16 所示，创建完成。

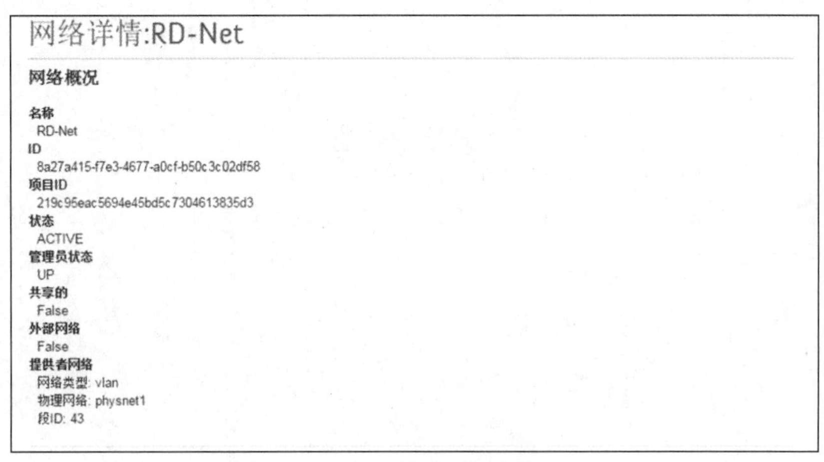

图 5-15　创建网络结果

图 5-16　创建子网结果

（2）通过 Shell 命令行创建网络和子网

① 为业务部通过命令创建网络和子网，命令如下，执行结果如下。

```
# neutron net-create BS-Net1Q
Created a new network:
+---------------------------+--------------------------------------+
| Field                     | Value                                |
+---------------------------+--------------------------------------+
| admin_state_up            | True                                 |
| id                        | ad1577b8-4307-4c40-a84d-6f4b0ebe55e5 |
| name                      | BS-Net1Q                             |
| provider:network_type     | vlan                                 |
| provider:physical_network | physnet1                             |
| provider:segmentation_id  | 45                                   |
| shared                    | False                                |
| status                    | ACTIVE                               |
| subnets                   |                                      |
| tenant_id                 | a3c430debe7e48d4a2f58c96dd89746b     |
+---------------------------+--------------------------------------+
# neutron subnet-create --name BS-Subnet --gateway 172.24.4.1 --allocation-pool
start=172.24.4.2,end=172.24.4.254 BS-Net 172.24.4.0/24
Created a new subnet:
+-------------------+------------------------------------------------+
| Field             | Value                                          |
+-------------------+------------------------------------------------+
| allocation_pools  | {"start": "172.24.4.2", "end": "172.24.4.254"} |
| cidr              | 172.24.4.0/24                                  |
| dns_nameservers   |                                                |
| enable_dhcp       | True                                           |
| gateway_ip        | 172.24.4.1                                     |
| host_routes       |                                                |
| id                | 36566e2f-2d32-4fb6-aed7-a2255b370e0f           |
| ip_version        | 4                                              |
| name              | BS-Subnet                                      |
| network_id        | ad1577b8-4307-4c40-a84d-6f4b0ebe55e5           |
| tenant_id         | a3c430debe7e48d4a2f58c96dd89746b               |
+-------------------+------------------------------------------------+
```

② 为 IT 工程部通过命令创建网络和子网，命令如下，执行结果如下。

```
# neutron net-create IT-Net
Created a new network:
+---------------------------+--------------------------------------+
| Field                     | Value                                |
+---------------------------+--------------------------------------+
| admin_state_up            | True                                 |
| id                        | ff07e63d-6d52-4262-bc46-578035635d43 |
| name                      | IT-Net                               |
| provider:network_type     | vlan                                 |
| provider:physical_network | physnet1                             |
| provider:segmentation_id  | 45                                   |
+---------------------------+--------------------------------------+
```

shared	False
status	ACTIVE
subnets	
tenant_id	a3c430debe7e48d4a2f58c96dd89746b

Field	Value
admin_state_up	True
id	ff07e63d-6d52-4262-bc46-578035635d43
name	IT-Net
provider:network_type	vlan
provider:physical_network	physnet1
provider:segmentation_id	45
shared	False
status	ACTIVE
subnets	
tenant_id	a3c430debe7e48d4a2f58c96dd89746b

```
# neutron subnet-create --name IT-Subnet --gateway 172.24.5.1 --allocation-pool
start=172.24.5.2,end=172.24.5.254 IT-Net 172.24.5.0/24
Created a new subnet:
```

Field	Value
allocation_pools	{"start": "172.24.5.2", "end": "172.24.5.254"}
cidr	172.24.5.0/24
dns_nameservers	
enable_dhcp	True
gateway_ip	172.24.5.1
host_routes	
id	1a945f77-cbd7-40d9-befb-d46e4f5eac3c
ip_version	4
name	IT-Subnet
network_id	ff07e63d-6d52-4262-bc46-578035635d43
tenant_id	a3c430debe7e48d4a2f58c96dd89746b

（3）创建外来访问使用网络

创建外来访问使用网络的具体命令如下，执行结果如下。

```
# neutron net-create Guest-Net
Created a new network:
```

Field	Value
admin_state_up	True
id	f6bb5ce7-47b0-4f67-88ca-d2cbd34e007f
name	Guest-Net
provider:network_type	vlan

```
| provider:physical_network | physnet1                         |
| provider:segmentation_id  | 45                               |
| shared                    | False                            |
| status                    | ACTIVE                           |
| subnets                   |                                  |
| tenant_id                 | a3c430debe7e48d4a2f58c96dd89746b |
# neutron subnet-create --name Guest-Subnet --gateway 172.24.6.1 --allocation-pool start=172.24.6.2,end=172.24.6.254 Guest-Net 172.24.6.0/24
Created a new subnet:
| Field             | Value                                          |
| allocation_pools  | {"start": "172.24.6.2", "end": "172.24.6.254"} |
| cidr              | 172.24.6.0/24                                  |
| dns_nameservers   |                                                |
| enable_dhcp       | True                                           |
| gateway_ip        | 172.24.6.1                                     |
| host_routes       |                                                |
| id                | 756d67e2-7a80-482f-a7a1-709610a95f6e           |
| ip_version        | 4                                              |
| name              | Guest-Subnet                                   |
| network_id        | f6bb5ce7-47b0-4f67-88ca-d2cbd34e007f           |
| tenant_id         | a3c430debe7e48d4a2f58c96dd89746b               |
```

3. 网络隔离

通过"访问&安全"控制，可实现不同 Vlan 之间的网络隔离。要求项目研发部和业务部之间网络隔离，只允许访问外部网络，不允许相互通信，IT 工程部网络可以对所有网络进行管理。

（1）创建项目研发部安全组规则

① 进入 Dashboard 界面找到"项目"。

② 在下拉菜单中，选中"访问&安全"。

③ 选择"创建安全组"，输入"名称"和"描述"的信息，单击"创建安全组"按钮，如图 5-17 所示。

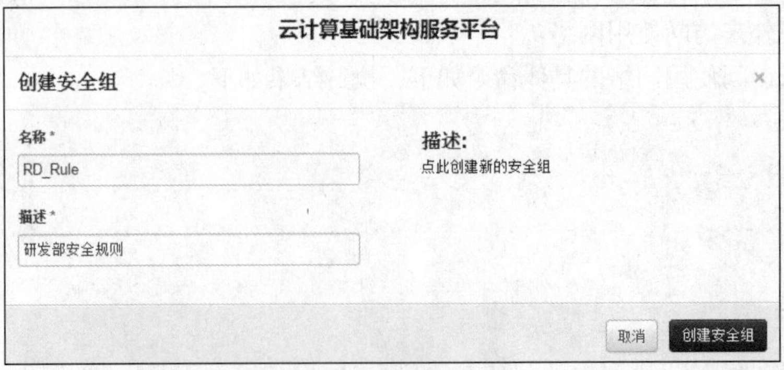

图 5-17 创建安全组

④ 在对应的安全组中，单击“管理规则”按钮，如图 5-18 所示。

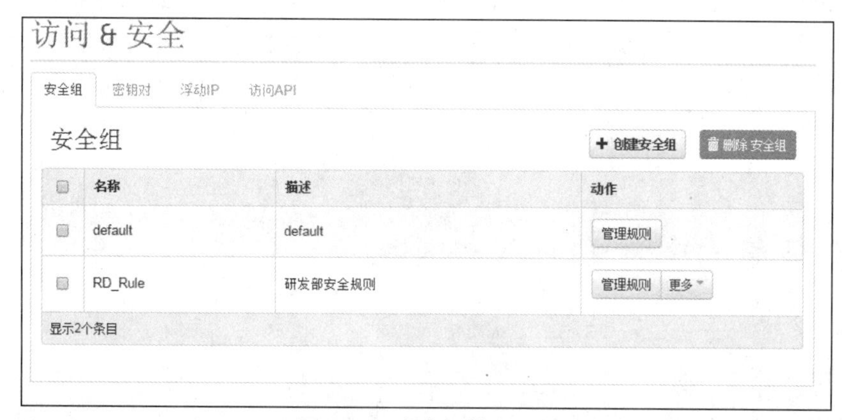

图 5-18　安全组管理界面

⑤ 单击“添加规则”按钮，在弹出的窗口中输入要添加的规则、方向、打开端口、端口、远程和 CIDR。

⑥ 首先添加出口规则，允许访问所有网络，添加入口规则 ALL ICMP、ALL TCP 和 ALL UDP，只允许项目研发部相互访问和 IT 工程部的访问，创建结果如图 5-19 所示。

管理安全组规则：RD_Rule

安全组规则　　+ 添加规则　删除规则

方向	输入类型	IP协议	端口范围	远程	动作
入口	IPv4	UDP	1 - 65535	172.24.3.0/24 (CIDR)	删除规则
入口	IPv4	TCP	1 - 65535	172.24.5.0/24 (CIDR)	删除规则
入口	IPv4	ICMP	-	172.24.3.0/24 (CIDR)	删除规则
出口	IPv4	任何	-	0.0.0.0/0 (CIDR)	删除规则
入口	IPv4	UDP	1 - 65535	172.24.5.0/24 (CIDR)	删除规则
入口	IPv4	TCP	1 - 65535	172.24.3.0/24 (CIDR)	删除规则
入口	IPv4	ICMP	-	172.24.5.0/24 (CIDR)	删除规则
出口	IPv6	任何	-	::/0 (CIDR)	删除规则

显示8个条目

图 5-19　安全组规则管理

（2）通过 CLI 界面创建业务部安全组规则

添加入口规则 ALL ICMP、ALL TCP 和 ALL UDP，只允许业务部相互访问以及 IT 工程部的访问，创建过程的命令如下，执行结果如下。

```
# nova secgroup-create BS_Rule 业务部安全组规则
+--------------------------------------+---------+------------------+
| Id                                   | Name    | Description      |
+--------------------------------------+---------+------------------+
| 629c817c-1c96-4eae-8b0c-17399125f560 | BS_Rule | 业务部安全组规则 |
+--------------------------------------+---------+------------------+
# nova secgroup-add-rule BS_Rule ICMP -1 -1 172.24.4.0/24
```

IP Protocol	From Port	To Port	IP Range	Source Group
icmp	-1	-1	172.24.4.0/24	

```
# nova secgroup-add-rule BS_Rule TCP 1 65535 172.24.4.0/24
```

IP Protocol	From Port	To Port	IP Range	Source Group
tcp	1	65535	172.24.4.0/24	

```
# nova secgroup-add-rule BS_Rule UDP 1 65535 172.24.4.0/24
```

IP Protocol	From Port	To Port	IP Range	Source Group
udp	1	65535	172.24.4.0/24	

```
# nova secgroup-add-rule BS_Rule ICMP -1 -1 172.24.5.0/24
```

IP Protocol	From Port	To Port	IP Range	Source Group
icmp	-1	-1	172.24.5.0/24	

```
# nova secgroup-add-rule BS_Rule TCP 1 65535 172.24.5.0/24
```

IP Protocol	From Port	To Port	IP Range	Source Group
tcp	1	65535	172.24.5.0/24	

```
# nova secgroup-add-rule BS_Rule UDP 1 65535 172.24.5.0/24
```

IP Protocol	From Port	To Port	IP Range	Source Group
udp	1	65535	172.24.5.0/24	

```
# nova secgroup-list-rules BS_Rule
```

IP Protocol	From Port	To Port	IP Range	Source Group
tcp	1	65535	172.24.5.0/24	
udp	1	65535	172.24.5.0/24	
icmp	-1	-1	172.24.5.0/24	
udp	1	65535	172.24.4.0/24	
tcp	1	65535	172.24.4.0/24	
icmp	-1	-1	172.24.4.0/24	

（3）创建 IT 工程部安全组规则

执行命令如下，执行后结果如下所示。

```
# nova secgroup-create IT_Rule 工程部安全组规则
```

Id	Name	Description
9f3968af-ca0a-4ac7-800b-00852d4426af	IT_Rule	工程部安全组规则

```
# nova secgroup-add-rule IT_Rule ICMP -1 -1 172.24.5.0/24
```

IP Protocol	From Port	To Port	IP Range	Source Group
icmp	-1	-1	172.24.5.0/24	

```
# nova secgroup-add-rule IT_Rule TCP 1 65535 172.24.5.0/24
```

IP Protocol	From Port	To Port	IP Range	Source Group
tcp	1	65535	172.24.5.0/24	

```
# nova secgroup-add-rule IT_Rule UDP 1 65535 172.24.5.0/24
```

IP Protocol	From Port	To Port	IP Range	Source Group
udp	1	65535	172.24.5.0/24	

```
# nova secgroup-list-rules IT_Rule
```

IP Protocol	From Port	To Port	IP Range	Source Group
udp	1	65535	172.24.5.0/24	
tcp	1	65535	172.24.5.0/24	
icmp	-1	-1	172.24.5.0/24	

项目六 虚拟化服务

在1959年6月的国际信息处理联合会（International Federation for Information Processing, IFIP）上，英格兰计算机科学家克里斯托弗•斯特雷奇（Christopher Strachey）在其学术报告《大型高速计算机中的时间共享》（Time Sharing in Large Fast Computers）一文中最早提出“Virtualization”虚拟化的概念，这是虚拟化概念的第一次出现，如今虚拟化已成为云计算基础架构的基石，云计算的工作内容实质就是对虚拟化的资源进行管理、调度和分配。本项目将带领大家深入了解虚拟化技术。

学习目标

- 了解虚拟化技术中的相关概念。
- 了解主流虚拟化技术。
- 理解虚拟化的工作原理。
- 掌握KVM虚拟化技术的具体实现方法。
- 掌握KVM虚拟机启动、删除、控制和监控等基本操作。

任务 虚拟化操作

任务要求

虚拟化技术是云计算技术的基石。Virsh（Virtual Shell）是由一个名为Libvirt的软件提供的管理工具，提供更高级的管理虚拟机的能力。小李接下来要了解虚拟化技术的概念、主流虚拟化技术、KVM虚拟化技术的实现，并熟悉工具Virsh，利用该工具掌握启动、删除、控制和监控KVM中虚拟机的基本操作。

相关知识

1. 虚拟化架构介绍

目前，主要的x86虚拟化架构包括以下几种情形。

（1）容器模式

虚拟机运行在传统的操作系统上，创建一个独立的虚拟化实例，指向底层托管操作系统，被称为“操作系统虚拟化”。

（2）主机模式

虚拟机被当作一个应用程序运行在传统的操作系统上，虚拟化的应用程序提供一个全仿真的硬件实例。这种虚拟化架构也被称为“托管”Hypervisor。

（3）裸机模式

虚拟机直接运行在硬件上，由虚拟化软件提供全仿真的硬件环境。这种类型也叫作“裸金属”。

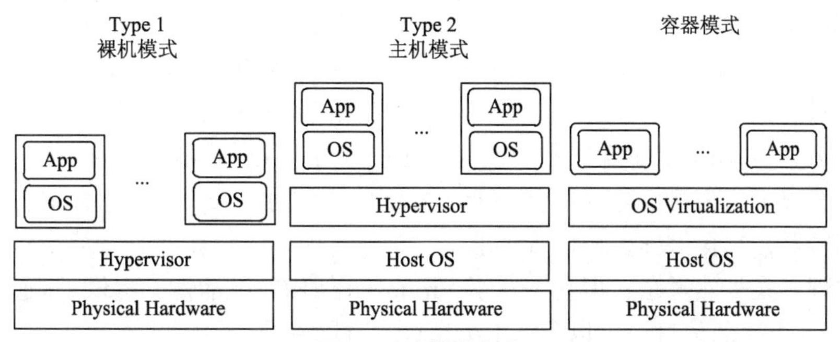

图 6-1　虚拟化架构

图 6-1 表示了 3 种虚拟化架构类型。在每种类型中，虚拟层都独立标注出来，以便清楚地表示它的逻辑位置。

除了上述的虚拟化架构之外，虚拟化的管理工具 Hypervisor 也起到了举足轻重的作用，目前针对其也有以下的基本概念。

虚拟机监视器（Virtual Machine Monitor，VMM），它主要负责创建、管理和删除虚拟化硬件。

半虚拟化（Paravirtualization），即需要修改软件，以便让它知道它运行在虚拟化环境中。对于特定的 Hypervisor，还包括下面一种或两种情形。

① 内核半虚拟化（Kernel Paravirtualization）：需要更改操作系统内核，同时需要 Guest OS 和 Hypervisor 兼容。

② 驱动半虚拟化（Driver Paravirtualization）：需要修改 Guest OS I/O 驱动（网络、存储等），如 VMware Tools 等。

2. 操作系统虚拟化

在容器模型中，虚拟层是通过创建虚拟操作系统实例实现的，它再指向根操作系统的关键系统文件，这些指针驻留在操作系统容器受保护的内存中，提供低内存开销，因此虚拟化实例的密度很大。密度是容器架构相对于Ⅰ型和Ⅱ型架构的关键优势之一，每个虚拟机都需要一个完整的客户机操作系统实例。

通过共享系统文件，所有容器可能只基于根操作系统就能够提供客户机。举一个简单的例子，一个基本的 Windows Server 2003 操作系统也可用于创建 Windows Server 2003 容器。同样，任何适用于根操作系统的系统文件的补丁和更新，其子容器也会继承，这提供了一个方便的维护方法。

但这也可能会造成损害，如果根操作系统受到破坏，客户机也会跟着被破坏。

在容器内，用户可以使用特定的应用程序、补丁 Fixes（但不是 Service Pack，它会更改操作系统 Kernel 的共享系统文件）和操作系统服务组件自定义客户机实例。对于那些在多数客户机容器中会使用到的服务或应用程序，它们所需要的功能应该被安装到根操作系统中，在客户机实例中使用类似于模板的方法自动获得这些功能。

在大多数情况下，容器的数量仅受宿主操作系统可用资源的限制。每个客户机都可能被配置为宿主机操作系统限制的最大硬件资源，这些可扩展特性与客户机管理的易用性，使容器方法成为需要高虚拟机密度的应用程序的很有实力的候选者，如虚拟桌面。

Parallels Virtuozzo 容器是当今业界领先的操作系统虚拟化产品，除了上述功能外，Virtuozzo 还提供了高可用和跨物理主机迁移客户机的功能（假设宿主机操作系统和补丁级别相同）。在架构上，Virtuozzo 实现了一个专有的内核服务抽象层（Kernal Service Abstract Layer，KSAL），以保护宿主操作系统文件，它在可写入文件系统上保存一份安全的副本，使单独修改客户机成为可能。与混合 Hypervisor 中的父分区类似，第一个虚拟实例是一个简单的管理容器，它提供虚拟机监视功能。

这种实现方式虽然复杂，但它展示了 Virtuozzo 架构的灵活性，提供了高虚拟机密度。关于 Virtuozzo 更多的信息，可以参考 Parallels 官方网站。

3. 托管

Type 2（托管）Hypervisor 通过软件层部署在操作系统上，被当作一个应用程序。和容器架构不同，Type 2 为虚拟机提供了一个完整、隔离和独立的运行环境，通常使用半虚拟化驱动来改善虚拟机网络和 I/O 的性能。但是，由于虚拟机必须通过宿主机操作系统才能访问硬件，所以性能比不上“裸金属”架构的虚拟机。另外，诸如高可用性的企业级特性和管理，在 Type 2 中是无法实现的。这些原因造成了 Type 2 通常用来做开发测试和桌面级别使用，而不用到企业级生产环境中。

常用的 Type 2 Hypervisor 包括 VMware Workstation、VirtualBox 和 Virtual Server 等。

4. 裸金属

Type 1（裸金属）Hypervisor 架构包括主要的企业级虚拟化产品，这种类型的 Hypervisor 直接运行在物理服务器上，为虚拟机提供最好的性能。通过 Intel 和 AMD CPU 的虚拟化指令集，Type 1 Hypervisor 能够让虚拟机在某些情况下，性能接近或者达到物理机的性能。

Type 1 架构有 3 个子分类。

（1）独立型（Stand-alone）

在独立型 Hypervisor 中，特别定制的 Hypervisor 可提供所有硬件虚拟化和 VMM 功能。这种架构体现在 VMware ESXi Server 产品线中。值得一提的是，VMware 不是一个基于 Linux 的 Hypervisor，而是一个经过定制化的非常复杂的 VMkernel，它提供虚拟机所有的监控和硬件虚拟化功能。图 6-2 所示是 VMware ESXi 的基本架构。在早期的 ESX 版本中，如 ESX 3 和 ESX 4，Hypervisor 还提供了一个 Linux 的服务控制台。

和大多数 Type 1 和 Type 2 Hypervisor 一样，VMware 需要在 Guest OS 上安装 VMware Tools 来使用网络和 I/O 驱动半虚拟化，以提高虚拟机性能。

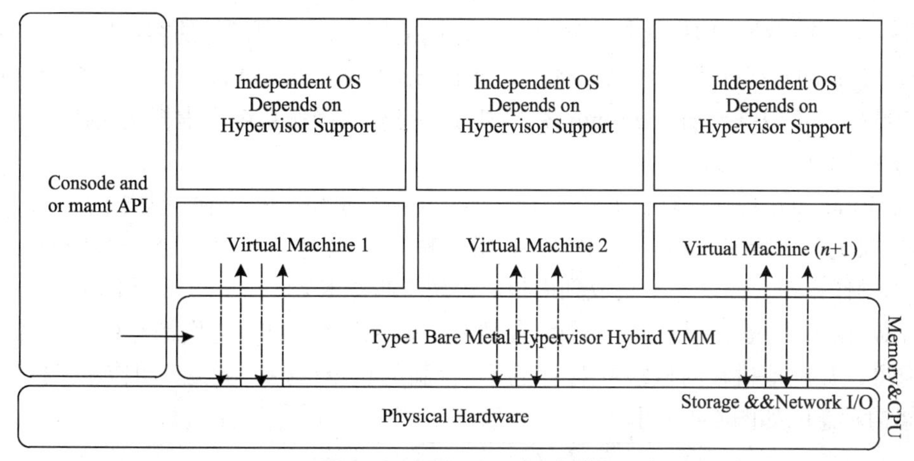

图 6-2　VMware ESXi 架构

（2）混合型（Hybird）

混合型架构包括一个软件模型，一个瘦 Hypervisor 联合一个父分区提供硬件虚拟化，它提供了虚拟机监视功能，这类模型主要包括微软的 Hyper-V 和基于 Xen 的 Hypervisor，代表产品为 Citrix XenServer、Microsoft Hyper-V。Hybird 架构示意图如图 6-3 所示。

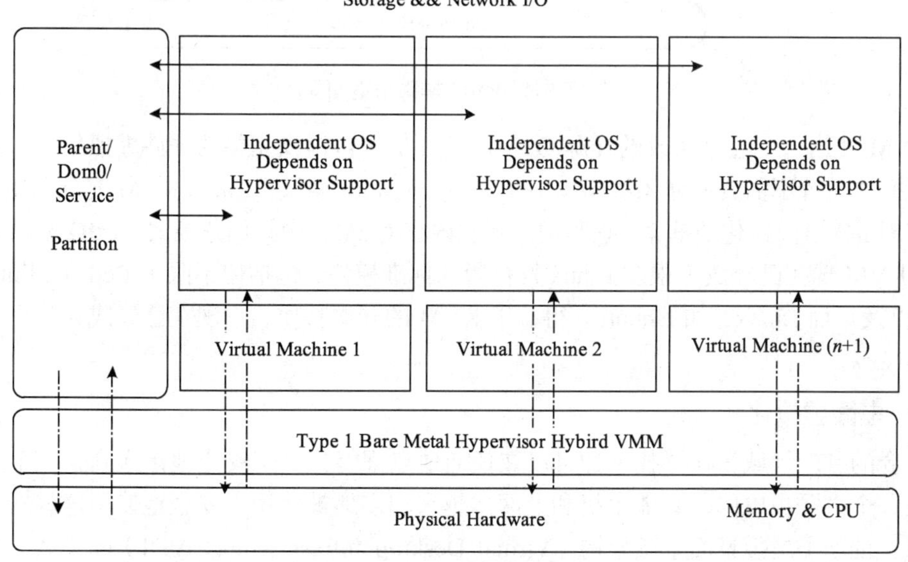

图 6-3　Hybird 架构示意图

父分区也叫作 Dom 0，它通常是一个运行在本地的完整的操作系统虚拟机，并具有根权限。例如，开启 Xen 在 Novell SUSE Linux Enterprise Server（SLES）上执行的 Dom 0，它将作为一个完整的 SLES 实例，提供虚拟机创建、修改、删除和其他类似配置任务的管理层。系统启动时，开启 Xen 的内核载入父分区，以 VMM 权限运行，作为 VM 管理的接口管理 I/O 堆栈。

与 VMware 类似，所有的混合型产品都为客户机提供了半虚拟化驱动，从而提高网络和 I/O 性能。不实现半虚拟化驱动的客户机必须遍历父分区的 I/O 堆栈，因此客户机的性能会下降。操作系统半虚拟化技术正变得越来越流行，以达到最佳的客户机性能，并改进跨 Hypervisor 的互操作性。例如，Microsoft Hyper-V/Windows Server 2008 R2 为 Windows Server 2008 和 SUSE Enterprise Linux 客户机提供完整的操作系统半虚拟化支持。

（3）混用型（Mixed）

基于 Linux 的内核虚拟机（KVM）Hypervisor 模型提供了一个独一无二的 Type 1 架构，它不是在裸机上执行 Hypervisor，KVM 利用开源 Linux（包括 RHEL、SUSE 和 Ubuntu 等）作为基础操作系统，提供一个集成到内核的模块（叫作 KVM），以实现硬件虚拟化，KVM 模块在用户模式下执行（与独立型和混合型 Hypervisor 不一样，它们都运行在内核/根模式下），但可以让虚拟机在内核级权限下使用一个新的指令执行上下文（叫作客户机模式）。KVM 架构示意图如图 6-4 所示。

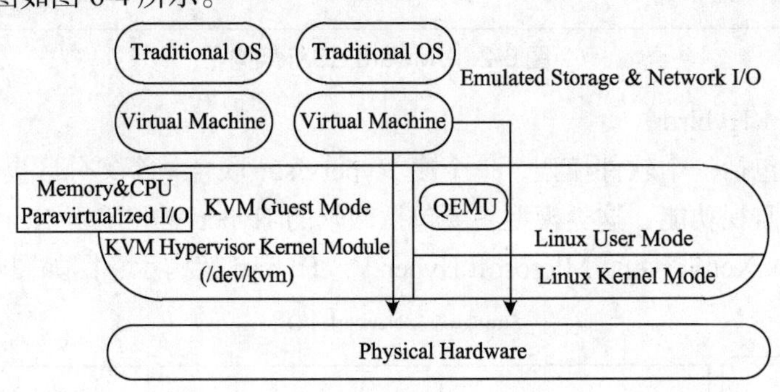

图 6-4　KVM 架构示意图

KVM 使用一个经过修改的开源 QEMU 硬件仿真包提供完整的硬件虚拟化，这意味着客户机操作系统不需要操作系统半虚拟化。与 VMware 类似，Linux KVM 充分利用 VirtIO 作为实现 I/O 半虚拟化的框架，它利用内置在内核 QEMU 中的用户模式 VirtIO 驱动来增强性能。KVM 现在已经成为很多 Linux 发行版的标准模块，包括但不限于 Red Hat Enterprise Linux 以及桌面 Linux，如 Ubuntu。得益于 KVM 的开源特质，它现在已经成为一个流行的 Hypervisor。

5. 桌面虚拟化

现今的 IT 领域正在寻找一种可以在任何地点都能让员工安全地接入到企业个人桌面的方案。个人平板电脑、智能手机和上网本越来越多地被使用，安全地使用这些新的设备的需求，推动了虚拟桌面基础架构（Virtual Desktop Infrastructure，VDI）的发展。除此之外，还有另外几个因素也促进着 VDI 的发展。

① 数据安全性和合规性。

② 管理现有传统桌面环境的复杂性和成本。

③ 不断增加的移动办公需求。

④ BYOD（Dring Your Own Device）的兴起。

⑤ 快速地恢复的能力。

VDI 是基于桌面集中的方式来给网络用户提供桌面环境的，这些用户使用设备上的远程显示协议（如 RDP、PCoIP 等）安全地访问他们的桌面。这些桌面资源被集中起来，允许用户在不同的地点进行访问。例如，在办公室打开一个 Word 应用，如果有事情临时出去了，在外地用平板电脑连接上虚拟桌面，则可以看到这个 Word 应用程序依然在桌面上，和离开办公室时的状态一致。这样使得系统管理员可以更好地控制和管理个人桌面，提高安全性。

集中化架构的想法早在大型机和终端客户机的年代就已经有了。在 20 世纪 90 年代，这种集中化架构被应用到服务器/客户端模式来满足用户的灵活性需求。这样的转变，促进了将计算处理集中到后台，而让用户可以将程序和文件保存在本地硬盘上的方式发展。真正的桌面虚拟化技术，是在服务器虚拟化技术成熟之后才出现的。

第一代桌面虚拟化技术，真正意义上将远程桌面的远程访问能力与虚拟操作系统结合起来，使得桌面虚拟化的企业应用成为可能。

首先，服务器虚拟化技术的成熟，以及服务器计算能力的增强，使得服务器可以提供多个桌面操作系统的计算能力。以当前 4 核双 CPU 的志强处理器、16GB 内存服务器为例，如果用户的 Windows XP 系统分配 256MB 内存，在平均水平下，一台服务器可以支撑 50～60 个桌面的运行，可以看到，如果将桌面集中使用虚拟桌面，那么 50～60 个桌面的采购成本将高于服务器的成本，而管理成本、安全因素还未被计算在内，所以服务器虚拟化技术的出现，使得企业大规模应用桌面虚拟化技术成为可能。

如果只是把台式机上运行的操作系统转变为服务器上运行的虚拟机，而用户无法访问，当然是不会被任何人接受的。所以，虚拟桌面的核心与关键，不是后台服务器虚拟化技术将桌面虚拟化，而是让用户通过各种手段，在任何时间、任何地点，通过任何可联网设备都能够访问到自己的桌面，即远程网络访问的能力。而这又转回到和应用虚拟化的共同点，即远程访问协议的高效性上。

提供桌面虚拟化解决方案的主要厂商包括微软、VMware 和 Citrix，而使用的远程访问协议主要有 3 种：第 1 种是早期由 Citrix 开发的，后来被微软购买并集成在 Windows 系统中的 RDP，这种协议被微软桌面虚拟化产品使用，而基于 VMware 的 Sun Ray 等硬件产品，也都是使用 RDP；第 2 种是 Citrix 自己开发的、独有的 ICA 协议，Citrix 将这种协议使用到其应用虚拟化产品与桌面虚拟化产品中；第 3 种是加拿大的 Teradici 公司开发的 PCoIP，它被使用到 VMware 的桌面虚拟化产品中，用于提供高质量的虚拟桌面用户体验。

协议效率决定了使用虚拟桌面的用户体验，而用户体验是决定桌面产品生命力的关键。从官方的文档与实际测试来看，在通常情况下，ICA 协议要优于 RDP 和 PCoIP，ICA 协议需要 30～40Kbps 的带宽，而 RDP 则需要 60Kbps，这些都不包括看视频、玩游戏以及 3D 制图状态下的带宽占用率。正是由于这个原因，虚拟桌面的用户体验有比较大的差别。一般情况下，在局域网环境中，一般的应用使用 RDP 和 ICA 协议都能正常运行，只不过是 RDP 会造成网络占用率较多的后果，但对于性能还不至于产生很大的影响；在广域网甚至互联网上，RDP 基本不可用；而在视频观看、Flash 播放和 3D 设计等应用上，即使在局域网中，RDP 的性能也会受到较大影响，但 ICA 协议的用户体验会很流畅。而且根据 Citrix 官方刚刚推出的 HDX 介绍，这方面的新技术会得到更快的推进。而微软和 VMware 也意识到了这一差别，微软转而加大对 RDP 的研发与优化，VMware 也和加拿大的 Teradici 公司合作使用其开发的 PCoIP，用于提供高质量的虚拟桌面用户体验。最新的 VMware View 5.0 产品提高了 PCoIP 的性能，并将带宽占用率降低了 75%，成为虚拟桌面的领跑协议。

第一代技术实现了远程操作和虚拟技术的结合，降低的成本使得虚拟桌面技术的普及成为可能，但是影响普及的并不仅仅是采购成本，管理成本和效率在这个过程中也是非常重要的一环。

纵观 IT 应用的历史，架构的变化和“三国”中的名言一样：分久必合，合久必分。从最早的主机-终端集中模式，到 PC 分布模式，再到今天的虚拟桌面模式，其实是计算使用权与管理权的博弈发展。开始时采用主机模式，可实现集中管理，但是应用困难，必须到机房去使用。PC 时代来临，所有计算都在 PC 上发生，但是 IT 的管理也变成分布式的，这也是 IT 部门的桌面管理员压力最大的原因，需要分布式管理所有用户的 PC，管理成本也大幅度上升。桌面虚拟化把用户操作环境与系统实际运行环境拆分，不必同时在一个位置，这样既满足了用户的灵活使用，同时也帮助 IT 部门实现了集中控制，从而解决了分布式带来的问题。但是，如果只是将 1 000 个员工的 PC 变成 1 000 个虚拟机，那么 IT 管理员的管理压力可能并没有降低，反而上升了，只不过是不用四处乱跑了而已。

为了提高管理性，第二代桌面虚拟化技术进一步将桌面系统的运行环境与安装环境拆分，应用与桌面拆分，配置文件拆分，从而大大降低了管理复杂度与成本，提高了管理效率。

用户可以针对下面的情况来简单计算一下。一个企业有 200 个用户，如果不进行拆分，那么 IT 管理员需要管理 200 个镜像（包含其中安装的应用与配置文件）。如果进行操作系统安装与应用还有配置文件的拆分，假设有 20 个应用，则使用应用虚拟化技术，不用在桌面安装应用，只是动态地将应用组装到桌面上，管理员只需要管理 20 个应用。配置文件也可以使用 Windows 内置的功能，将文件数据都保存在文件服务器上，这些信息不需要管理员管理，管理员只需要管理一台文件服务器就行。应用和配置文件的拆分，使得 200 个人的操作系统都是没有差别的 Windows XP，管理员只需要管理一个镜像（用这个镜像生成 200 个运行的虚拟操作系统，简单来讲，可以理解成类似于无盘工作站的模式）即可。所以，总而言之，IT 管理员只需要管理 20 个应用、1 台文件服务器和 1 个镜像，管理复杂性大大下降。

这种拆分也大大降低了对存储的需求量（少了 199 个 Windows XP 系统的存储），降低了采购和维护成本。更重要的是，在管理效率上，管理员只需要对一个镜像或者一个应用进行打补丁或者升级操作，所有的用户都会获得最新更新后的结果，从而提高了系统的安全性和稳定性，工作量也大大下降。

无论是 Hyper-V 还是 vShpere，它们都是封闭的软件，不对用户开放源代码，因此很难知道这些 Hypervisor 到底在做什么，如果要深入定制，形成用户自己的一套方案，则只能受制于厂商，乖乖地交不菲的授权费。随着开源市场的兴起，采用 KVM 作为服务器端的虚拟桌面方案的厂商也不断涌现，其中的代表有 Red Hat 和 Virtual Bridges。虽然和闭源软件相比，基于开源的虚拟桌面方案在功能上还有一定的差距，但与 RHEV 和 vShpere 相比，可以实现其 90%以上的功能，但是价格却低得多。基于社区的开源软件开发模式，可以快速推出新的功能，同时有实力的厂商的介入，会对开源软件的发展起到推动作用。

6. VDI 架构介绍

目前市面上的 VDI 方案，其基本架构如图 6-5 所示。

（1）用户访问层（User Access Layer）

用户访问层是用户进入 VDI 的入口。用户通过支持 VDI 访问协议的各种设备，如电脑、瘦客户端、上网本和手持移动设备等来进行访问。

（2）虚拟架构服务层（Virtual Infrastructure Service Layer）

虚拟架构服务层为用户提供安全、规范和高可用的桌面环境。用户访问层通过特定的显示协议和中间层实现该层通信，如 VMware 使用的是 RDP 和 PCoIP，Citrix 使用 HDX，Red Hat 使用 SPICE 等。

（3）存储服务层（Storage Service Layer）

存储服务层存储用户的个人数据、属性、母镜像和实际的虚拟桌面镜像。虚拟架构服务调用存储协议来访问数据。VDI 里面常用到的存储协议有 NFS（Network File System）、CIFS（Common Internet File System）、iSCSI 和 Fibre Channel 等。

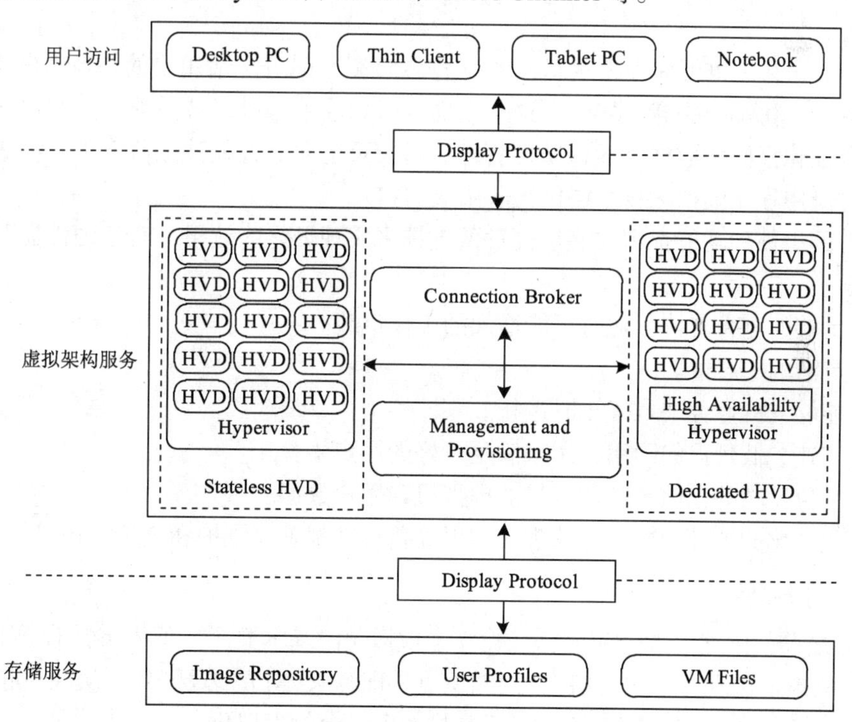

图 6-5　VDI 架构示意图

虚拟架构服务层有如下主要组件和功能。

（1）Hypervisor：为虚拟桌面的虚拟机提供虚拟化运行环境。这些虚拟机就叫作用户虚拟桌面。

（2）用户虚拟桌面（Hosted Virtual Desktop）：虚拟机里面运行的桌面操作系统和应用就是一个用户虚拟桌面。

（3）连接管理器（Connection Broker）：用户的访问设备通过连接管理器来请求虚拟桌面。它管理访问授权，确保只有合法的用户能够访问 VDI。一旦用户被授权，连接管理器就将用户请求定向到分配的虚拟桌面上。如果虚拟桌面不可用，那么连接管理器将从管理和提供服务（Management and Provisioning Service）中申请一个可用的虚拟桌面。

（4）管理和提供服务（Management and Provisioning Service）：管理和提供服务可集中化管理虚拟架构，它提供单一的控制界面来管理多项任务。它提供镜像管理、生命周期管理和监控虚拟桌面功能。

（5）高可用性服务（High Availability Service）：高可用性（HA）服务可保证虚拟机在关键的软件或者硬件出现故障时能够正常运行。HA 可以是连接管理器功能的一个部分，为无状态 HVD 提供服务，也可以为全状态 HVD 提供单独的故障转移服务。

有两种类型的 HVD 虚拟机分配模式：永久和非永久。

一个永久（也可称为全状态或者独占）HVD 是被分配给特定的用户的（类似于传统 PC 的形式）。用户每次登录时，连接的都是同一个虚拟机，用户在这个虚拟机上安装或者修改的应用和数据将被保存下来，用户注销后也不会丢失。这种独占模式非常适合需要自己安装更多的应用程序，将数据保存在本地，保留当前状态，以便下次登录后可以继续工作的情况。

一个非永久（也可称为无状态或者池）HVD 是临时分配给用户的。当用户注销后，所有对镜像的变化都被丢弃。接着，这个桌面进入到池中，可以被另外一个用户连接使用。用户的个性化桌面和数据将通过属性管理、目录重定向等方式被保留下来。特定的应用程序将通过应用程序虚拟化技术提供给非永久 HVD。

所有的组件和功能都需要底层硬件的支撑来实现最终的效果，对于硬件需要做到如下几个方面。

① 足够的电力支持，随时可能增加的工作负载。

② 灵活的扩展性。

③ 能够支持业务 7×24 小时运行。

④ 高速、低延迟的网络，用来满足更好的用户体验。

⑤ 性价比高的存储来存放大量的虚拟机和用户数据。

⑥ 集中化的硬件和虚拟化管理，用以简化自动部署、维护和支持工作。

7．虚拟化原理

通过之前的阐述，大家应该已经建立了虚拟化的基本概念。下面将针对当下最流行的 x86 平台的虚拟化进行详述。这部分的知识也是理解 KVM 的必要准备。从现在起，使用“虚拟化”一词时，如果没有特殊说明，都是指 x86 平台的虚拟化。

处于虚拟机底层的是整个物理系统，也就是我们平常看得见、摸得着的系统硬件，主要包括处理器、内存和 I/O 设备（这一点相信有主机 DIY 经验的同学是非常熟悉的）。在物理系统之上，与以往熟悉的操作系统模型不同，运行的是虚拟机监控器（缩写为 VMM 或 Hypervisor）。虚拟机监控器的主要职能是管理真实的物理硬件平台，并为每个虚拟客户机提供对应的虚拟硬件平台。图 6-6 中绘制了 3 个虚拟机的实例，每个虚拟机看起来就像是一个小的但是完整的计算机系统，具有自己的“系统硬件”，包括自己的处理器、内存和 I/O 设备。在这个计算机系统上，运行着虚拟机自己的操作系统，如 Linux 和 Windows。

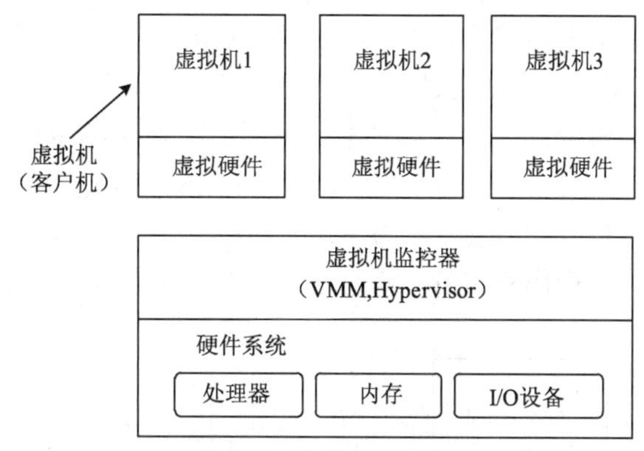

图 6-6　虚拟化模型

既然每一个虚拟机看起来就是一个小的“真实”计算机系统，那里面可以运行自己的虚拟机监控器吗？答案是肯定的。这种情况一般称为“嵌套虚拟化”。KVM 支持嵌套虚拟化技术，只是嵌套虚拟化的实现还远远没有达到很稳定和成熟的状态。嵌套虚拟化并不是本章要讲述的内容。

一个 x86 平台的核心是其中的处理器，处理器运行程序代码，访问内存和 I/O 设备。所以，x86 平台虚拟化技术的核心部分是处理器的虚拟化。只要处理器虚拟化技术支持“截获并重定向”，内存和 I/O 设备的虚拟化都可以基于处理器虚拟化技术实现。在处理器虚拟化技术的基础上，为了增强虚拟机的性能，内存虚拟化和 I/O 虚拟化的新技术也不断被加入到 x86 平台虚拟化技术中。x86 平台虚拟化技术从单一的处理器开始，逐步涉及芯片组、网卡、存储设备以及 GPU 的虚拟化。

（1）KVM 原理

① KVM 架构。

了解了虚拟化的基本模型之后，接下来，我们来看一下 KVM 的具体架构。从虚拟机的基本架构上来区分，虚拟机一般分为两种，我们称之为“类型一”和“类型二”。

其中，“类型一”虚拟机在系统加电之后首先加载运行虚拟机监控程序，而传统的操作系统则运行在其创建的虚拟机中。“类型一”虚拟机的监控程序，从某种意义上说，可以视为一个特别为虚拟机而优化裁剪的操作系统内核。因为，虚拟机监控程序作为运行在底层的软件层，必须实现诸如系统的初始化、物理资源的管理等操作系统的功能，它对虚拟机的创建、调度和管理，与操作系统对进程的创建、调度和管理有共同之处。这一类型的虚拟机监控程序一般会提供一个具有一定特权的特殊虚拟机，由这个特殊虚拟机来运行需要提供给用户日常操作和管理使用的操作系统环境。著名的开源虚拟化软件 Xen、商业软件 VMware ESX/ESXi 和微软的 Hyper-V 就是“类型一”虚拟机的代表。

与“类型一”虚拟机的方式不同，“类型二”虚拟机的监控程序，在系统加电之后仍然运行一般意义上的操作系统（也就是俗称的宿主机操作系统）。虚拟机监控程序作为特殊的应用程序，可以视作操作系统功能的扩展。对于“类型二”虚拟机来说，其最大的优势在于可以充分利用现有的操作系统。因为虚拟机监控程序通常不必自己实现物理资源的管理和调度算法，所以实现起来比较简洁。但是，这一类型的虚拟机监控程序既然依赖操作系统来实现管理和调度，就同样会受到宿主操作系统的一些限制。例如，通常仅仅为了虚拟机的优化，而对操作系统做出修改。本书的主角 KVM 就是属于“类型二”虚拟机，另外，VMware Workstation、VirtualBox 也是属于“类型二”虚拟机。

了解了基本的虚拟机架构之后，再来看一下图 6-7 所示的 KVM 的基本架构。显而易见，KVM 是一个基于宿主操作系统的“类型二”虚拟机。在这里，可以再一次看到实用至上的 Linux 设计哲学，既然“类型二”虚拟机是最简洁和容易实现的虚拟机监控程序，那么通过内核模块的形式实现出来就好。其他的部分尽可能充分利用 Linux 内核的既有实现，最大限度地重用代码。

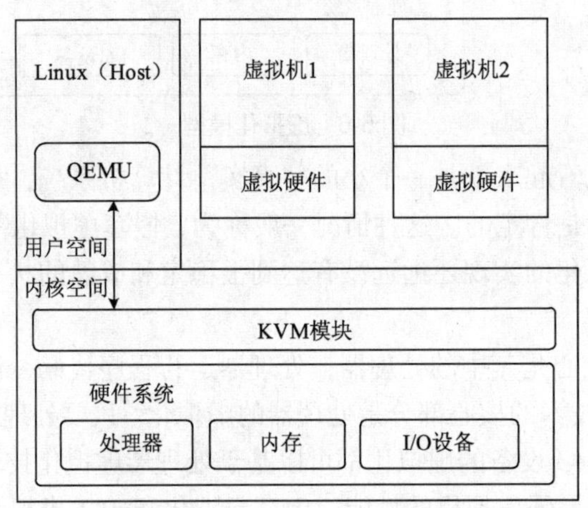

图 6-7　KVM 的基本架构

在图 6-7 中，左侧是一个标准的 Linux 操作系统，可以是 RHEL、Fedora 或 Ubuntu 等。KVM 内核模块在运行时按需加载进入内核空间运行。KVM 本身不执行任何设备模拟，需要用户空间程序 QEMU 通过/dev/kvm 接口设置一个虚拟客户机的地址空间，向它提供模拟的 I/O 设备，并将它的视频显示映射回宿主机的显示屏。

② KVM 模块。

KVM 模块是 KVM 虚拟机的核心部分，其主要功能是初始化 CPU 硬件，打开虚拟化模式，然后将虚拟客户机运行在虚拟机模式下，并对虚拟客户机的运行提供一定的支持。

为了软件的简洁和性能，KVM 模块仅支持硬件虚拟化。自然而然，打开并初始化系统硬件以支持虚拟机的运行，是 KVM 模块的职责所在。以 KVM 模块在 Intel 公司的 CPU 上运行为例，在被内核加载的时候，KVM 模块会先初始化内部的数据结构；做好准备之后，KVM 模块检测系统当前的 CPU，然后打开 CPU 控制寄存器 CR4 中的虚拟化模式开关，并通过执行 VMXON 指令将宿主操作系统（包括 KVM 模块本身）置于虚拟化模式中的根模式；最后，KVM 模块创建特殊设备文件/dev/kvm 并等待来自用户空间的命令。接下来，虚拟机的创建和运行将是一个用户空间的应用程序（QEMU）和 KVM 模块相互配合的过程。

KVM 模块与用户空间 QEMU 的通信接口主要是一系列针对特殊设备文件的 IOCTL 调用，如上所述，KVM 模块加载之初，只存在/dev/kvm 文件，而针对该文件的最重要的 IOCTL 调用就是“创建虚拟机”。在这里，“创建虚拟机”可以理解成 KVM 模块为了某个特定的虚拟客户机（用户空间程序创建并初始化）创建对应的内核数据结构。同时，KVM 模块还会返回一个文件句柄来代表所创建的虚拟机。针对该文件句柄的 IOCTL（控制代码）调用可以对虚拟机做相应的管理，例如创建用户空间虚拟地址和客户机物理地址及真实内存物理地址的映射关系，再例如创建多个可供运行的虚拟处理器（VCPU）。同样，KVM 模块会为每一个创建出来的虚拟处理器生成对应的文件句柄，对虚拟处理器相应的文件句柄进行相应的 IOCTL 调用，从而对虚拟处理器进行管理。

针对虚拟处理器的最重要的 IOCTL 调用就是"执行虚拟处理器"。通过它，用户空间准备好的虚拟机在 KVM 模块的支持下，被置于虚拟化模式中的非根模式下，开始执行二进制指令。在非根模式下，所有敏感的二进制指令都会被处理器捕捉到，处理器在保存现场之后自动切换到根模式，由 KVM 模块决定如何进一步处理（要么由 KVM 模块直接处理，要么返回用户空间交由用户空间程序处理）。

除了处理器的虚拟化，内存虚拟化也是由 KVM 模块实现的。实际上，内存虚拟化往往是一个虚拟机实现中代码量最大、实现最复杂的部分（至少，在硬件支持二维地址翻译之前是这样的）。众所周知，处理器中的内存管理单元（MMU）是通过页表的形式将程序运行的虚拟地址转换成为物理内存地址的。在虚拟机模式下，内存管理单元的页表则必须在一次查询的时候完成两次地址转换。这是因为，除了要将客户机程序的虚拟地址转换成为客户机物理地址以外，还必须将客户机物理地址转换成为真实物理地址。KVM 模块使用了影子页表的技术来解决这个问题，在客户机运行的时候，处理器真正使用的页表并不是客户机操作系统维护的页表，而是 KVM 模块根据这个页表维护的另外一套影子页表。影子页表的机制比较复杂，感兴趣的同学可以自行翻阅相关材料，这里不再展开。

影子页表实现复杂，而且有时候开销很大。为了解决这个问题，新的处理器在硬件上做了增强（Intel 的 EPT 技术）。通过引入第二级页表来描述客户机虚拟地址和真实物理地址的转换，硬件可以自动进行两级转换生成正确的内存访问地址。KVM 模块将其称为二维分页机制。

处理器对设备的访问主要是通过 I/O 指令和 MMIO 来完成的，其中，I/O 指令会被处理器直接截获，MMIO 会通过配置内存虚拟化来捕捉。但是，外设的模拟一般并不由 KVM 模块负责。一般来说，只有对性能要求比较高的虚拟设备才会由 KVM 内核模块直接负责，如虚拟中断控制器和虚拟时钟，这样可以大大减少处理器的模式切换的开销。大部分的 I/O 设备还是会交给用户态程序 QEMU 来负责。

（2）QEMU 模型

QEMU 本身并不是 KVM 的一部分，其自身就是一个著名的开源虚拟机软件。与 KVM 不同，QEMU 虚拟机是一个纯软件的实现，所以性能低下。但是，其优点是在支持 QEMU 本身编译运行的平台上就可以实现虚拟机的功能，甚至虚拟机可以与宿主机不是同一个架构。作为一个存在已久的虚拟机，QEMU 中有整套的虚拟机实现代码，包括处理器虚拟化、内存虚拟化，以及 KVM 使用到的虚拟设备模拟（如网卡、显卡、存储控制器和硬盘等）。

为了简化开发和代码重用。KVM 在 QEMU 的基础上进行了修改。虚拟机运行期间，QEMU 会通过 KVM 模块提供的系统调用进入内核，由 KVM 模块负责将虚拟机置于处理器的特殊模式下运行。当虚拟机进行 I/O 操作时，KVM 模块会从上次的系统调用出口处返回 QEMU，由 QEMU 来负责解析和模拟这些设备。

从 QEMU 角度来看，也可以说 QEMU 使用了 KVM 模块的虚拟化功能，为自己的虚拟机提供硬件虚拟化的加速，从而极大地提高了虚拟机的性能。除此之外，虚拟机的配置和创建，虚拟机运行依赖的虚拟设备，虚拟机运行时的用户操作环境和交互，以及一些针对虚拟机的特殊技术（诸如动态迁移），都是由 QEMU 自己实现的。

从 QEMU 和 KVM 模块之间的关系可以看出，这是典型的开源社区在代码共用和开发项目共用上的合作。诚然，QEMU 可以选择其他的虚拟机或技术来加速，如 Xen 或者 KQEMU；KVM 也可以选择其他的用户空间程序作为虚拟机实现，只要它按照 KVM 提供的 API 来设计。但是在现实中，QEMU 与 KVM 两者的结合是最成熟的选择，这对一个新开发和后起的项目（KVM）来说，无疑多了一份未来成功的保障。

（3）Libvirt

QEMU-KVM 工具可以创建和管理 KVM 虚拟机，Red Hat 为 KVM 开发了更多的辅助工具，如 Libvirt、Libguestfs 等。原因是 QEMU 工具效率不高，不易于使用。Libvirt 是一套提供了多种语言接口的 API，为各种虚拟化工具提供一套方便、可靠的编程接口，不仅支持 KVM，而且支持 Xen 等其他虚拟机。使用 Libvirt 时，只需要通过 Libvirt 提供的函数连接到 KVM 或 Xen 宿主机，便可以用同样的命令控制不同的虚拟机了。Libvirt 不仅提供了 API，还自带一套基于文本的管理虚拟机的命令 Virsh，可以通过使用“Virsh”命令来使用 Libvirt 的全部功能。但最终用户更渴望的是图形用户界面，这就是 Virt Manager。它是一套用 Python 编写的虚拟机管理图形界面，用户可以通过它直观地操作不同的虚拟机。Virt Manager 就是利用 Libvirt 的 API 实现的。Libvirt 是一个软件集合，便于使用者管理虚拟机和其他虚拟化功能，如存储和网络接口管理等。Libvirt 概括起来包括一个 API 库、一个 Daemon（Libvirtd）和一个命令行工具（Virsh）。

Libvirt 的主要目标是提供一种单一的方式管理多种不同的虚拟化提供方式和 Hypervisor，Libvirt 的工作原理如图 6-8 所示。

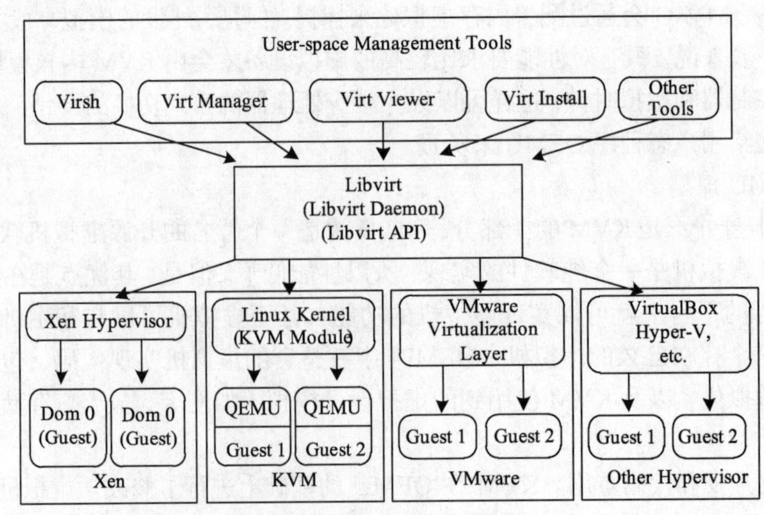

图 6-8 Libvirt 的工作原理

管理工具 Virsh、Virt Manager、Virt Install 等是通过使用 Libvirt 提供的 API，对虚拟化程序（Hypervisor）在各物理节点（Node）上虚拟化出的多个域（Domain，客户机操作系统 Guest OS）进行操作管理的。

Libvirt 的主要功能如下。

① 虚拟机管理：包括不同的领域生命周期操作，如启动、停止、暂停、保存、恢复和迁移。

支持多种设备类型的热插拔操作，包括磁盘、网卡、内存和 CPU。

② 远程机器支持：只要在机器上运行了 Libvirt Daemon，包括远程机器，所有的 Libvirt 功能就都可以访问和使用。

支持多种网络远程传输，可以使用最简单的 SSH，不需要额外配置工作。例如，example.com 运行了 Libvirt，而且允许 SSH 访问，下面的命令行就可以在远程的主机上使用 Virsh 命令行。

③ 存储管理：任何运行了 Libvirt Daemon 的主机都可以用来管理不同类型的存储，可以创建不同格式的文件映像（QCOW2、VMDK、RAW 等），挂接 NFS 共享，列出现有的 LVM 卷组，创建新的 LVM 卷组和逻辑卷。对未处理过的磁盘设备分区，挂接 iSCSI 共享等。因为 Libvirt 可以远程工作，所有这些都可以通过远程主机使用。

④ 网络接口管理：任何运行了 Libvirt Daemon 的主机都可以用来管理物理和逻辑的网络接口。现有的接口卡，配置、创建接口，以及桥接、VLAN 和关联设备等，通过 Netcf 都可以支持。任何运行了 Libvirt Daemon 的主机都可以用来管理和创建虚拟网络。Libvirt 虚拟网络使用防火墙规则作为路由器，让虚拟机可以透明访问主机的网络。

任务实现

1. 使用 KVM 管理工具

Virsh（Virtual Shell）是由一个名为 Libvirt 的软件提供的管理工具，提供更高级的管理虚拟机的能力。Virsh 的大部分功能与 XM 一样，可以利用 Virsh 来启动、删除、控制和监控 KVM 中所有的虚拟机。

（1）Virsh 的安装

安装环境为 CentOS 6.5 操作系统，硬件需要支持虚拟化。

① 安装 KVMLi bvirted。

```
# yum install kvm kmod-kvm qemu kvm-qemu-img virt-viewer virt-manager libvirt
libvirt-python python-virtinst
```

② 启动服务。

```
# service messagebus start
# service haldaemon start
# service libvirtd start
# chkconfig messagebus on
# chkconfig haldaemon on
# chkconfig libvirtd on
```

③ 检查 KVM 是否安装成功，执行结果如下。

```
# lsmod | grep kvm
kvm_intel              54285  0
kvm                    333172  1 kvm_intel
```

Virsh 提供两种执行模式：直接模式（Direct Mode）与互动模式（Interactive Mode）。在直接模式里，必须在 Shell 中以参数、自变量的方式来执行 Virsh。在互动模式中，Virsh 会提供一个提示字符串，在该提示字符串后，可以输入要执行的命令。如果执行 Virsh 时没有指定任何参数或自变量，则默认进入互动模式。

（2）Virsh 的操作

Virsh 的操作命令如表 6-1 所示。

表 6-1 Virsh 的操作命令

命令	说明
help	显示该命令的说明
quit	结束
virsh	获得一个特殊的 Shell
connect	连接到指定的虚拟机服务器
create	启动一个新的虚拟机
destroy	删除一个虚拟机
start	开启（已定义的）非启动的虚拟机
define	定义一个虚拟机
undefine	取消定义的虚拟机
dumpxml	转储虚拟机的设置值
list	列出虚拟机
reboot	重新启动虚拟机
save	存储虚拟机的状态
restore	回复虚拟机的状态
suspend	暂停虚拟机的执行
resume	继续执行该虚拟机
dump	将虚拟机的内核转储到指定的文件，以便进行分析与排错
shutdown	关闭虚拟机
setmem	修改内存的大小
setmaxmem	设置内存的最大值
setvcpus	修改虚拟处理器的数量

（3）测试安装 CentOS 6.5

① 创建虚拟机的磁盘文件，执行结果如下。

```
# qemu-img create -f raw demo.img 10G
Formatting 'demo.img', fmt=raw size=10737418240
```

② 通过 QEMU-KVM 创建虚拟机，执行结果如下。

```
 # virt-install --virt-type kvm --name centos --ram 1024 --cdrom CentOS-6.5-
x86_64-bin_DVD.iso --disk demo.img,format=raw  --graphics vnc, listen=0.0.0.0
-noautoconsole
Starting install...
Creating domain...                                 |    0 B     00:00
Domain installation still in progress. You can reconnect to
the console to complete the installation process.
```

③ 通过 VNC 客户端安装 CentOS。

通过“Virsh”命令查看虚拟机 VNC 的端口号，执行结果如下。

```
# virsh vncdisplay centos
:0
```

通过 VNC 连接虚拟机，如图 6-9 所示。连接成功后即可按照之前的步骤安装操作系统。

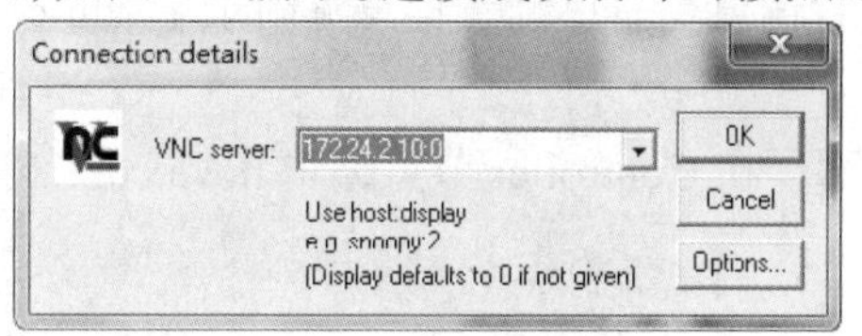

图 6-9　连接虚拟机

（4）通过“Virsh”命令查看虚拟机列表

```
# virsh list
```

执行结果如下。

```
Id      Name                        State
-----------------------------------------------------------
 4           centos                        running
```

通过“Virsh”命令强制关闭虚拟机。

```
# virsh destroy centos
Domain centos destroyed

# virsh list --all
```

执行结果如下。

```
Id      Name                        State
-----------------------------------------------------------
 -           centos                        shut off
```

通过“Virsh”命令删除虚拟机。

```
# virsh undefine centos
Domain centos has been undefined
```

执行结果如下。

```
[root@c1-controller opt]# virsh list --all
Id      Name                              State
----------------------------------------------------
```

实际上，在 KVM 中会把虚拟机的配置写到一个格式为.xml 的配置文件中，可根据实际情况通过“virsh edit domain”命令来对虚拟机的配置文件进行修改，感兴趣的同学可以研究一下，在此不做讲解。

2. 具体任务操作

利用 Virsh 工具管理实例，查看实例的运行情况和具体配置信息。

```
# virsh list --all -all
Id      Name                              State
----------------------------------------------------
3    instance-00000003              running
4    instance-00000004              running
```

查看 KVM 虚拟机配置信息（具体配置信息见附录十 vm-conf.txt），配置第三方 VNC 组件，如 novnc、spice。

采用 novnc 的虚拟桌面，在 Controller 节点编辑 nova.conf 文件，添加以下参数。

```
# vi /etc/nova/nova.conf
my_ip = 172.24.2.10
vncserver_listen = 172.24.2.10
vncserver_proxyclient_address = 172.24.2.10
```

重启 Nova 相关服务。

```
# service openstack-nova-api restart
# service openstack-nova-cert restart
# service openstack-nova-consoleauth restart
# service openstack-nova-scheduler restart
# service openstack-nova-conductor restart
# service openstack-nova-novncproxy restart
```

在 Compute 节点编辑 nova.conf 文件，添加以下参数。

```
# vi /etc/nova/nova.conf
my_ip = compute
vnc_enabled = True
vncserver_listen = 0.0.0.0
vncserver_proxyclient_address = compute
novncproxy_base_url = http://172.24.2.10:6080/vnc_auto.html
```

重启 Nova 相关服务。

```
# service openstack-nova-compute restart
```

进入 Dashboard 界面找到“项目”选项。

打开 compute 面板，选择实例。

找到相应实例，单击“更多”按钮，打开控制台进行验证，如图 6-10 所示。

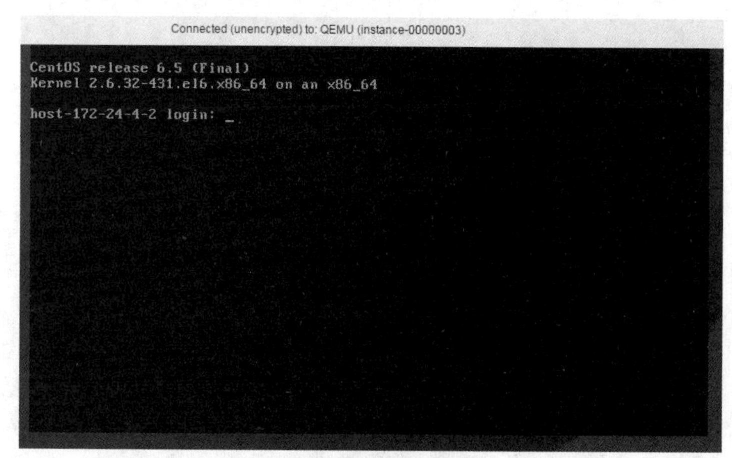

图 6-10　控制台验证

采用 spice 的虚拟桌面，在 Controller 节点编辑 nova.conf 文件，添加以下参数。

```
# vi /etc/nova/nova.conf
[spice]

enabled = True
html5proxy_base_url = http://172.24.2.10:6082/spice_auto.html
agent_enabled = True
keymap = en-us
server_listen = 0.0.0.0
server_proxyclient_address = 172.24.2.10
```

重启 Nova 相关服务。

```
# service openstack-nova-spicehtml5proxy restart
# service openstack-nova-novncproxy  stop
# service openstack-nova-scheduler restart
# service openstack-nova-api restart
# service openstack-nova-cert restart
# service openstack-nova-consoleauth restart
# service openstack-nova-scheduler restart
# service openstack-nova-conductor restart
# service openstack-nova-novncproxy stop
```

在 Compute 节点编辑 nova.conf 文件，添加以下参数。

```
# vi /etc/nova/nova.conf
```

```
[DEFAULT]
auth_strategy = keystone
rpc_backend = qpid
qpid_hostname = controller
my_ip = compute
vnc_enabled = False
vncserver_listen = 0.0.0.0
vncserver_proxyclient_address = compute
#novncproxy_base_url = http://172.24.2.10:6080/vnc_auto.html
[spice]
enabled = True
html5proxy_base_url = http://172.24.2.10:6082/spice_auto.html
agent_enabled = False
keymap = en-us
server_listen = 0.0.0.0
server_proxyclient_address = 172.24.2.20
```

重启 Nova 相关服务。

```
# service openstack-nova-compute restart
```

打开 compute 面板，选择实例。

找到相应实例，单击“更多”按钮，打开控制台进行验证。

提示

当控制台由 novnc 的模式改为 spice 的模式后，原有的虚拟机采用的控制台模式不会改变，新建的实例会使用 spice 模式。如需使用原有虚拟机的控制台，则使用“virsh edit instance”命令来更改控制台模式，将“graphics type”该成 spice 模式即可。

项目七 存储服务

在 OpenStack 中，有 3 个与存储相关的组件：Swift——提供对象存储（Object Storage），Glance——提供虚拟机镜像存储和管理，Cinder——提供块存储（Block Storage）。Glance 在前面的项目中已经介绍，本项目将介绍其他 OpenStack 的存储服务，了解内部的存储机制和流程，同时使用存储为平台的其他服务提供后端存储和数据的安全备份。

学习目标

- 了解 Cinder、Swift 的概念及用法。
- 理解云平台中的 Cinder 云硬盘服务。
- 理解 Swift 内部的运行算法和备份机制。
- 掌握利用 Cinder 为虚拟机提供硬盘空间。
- 掌握利用 Swift 为虚拟机提供硬盘空间。

任务一 块存储服务

任务要求

块存储可将磁盘整个映射给主机使用，它能提供逻辑卷功能。本任务主要针对 OpenStack 的块存储服务进行剖析，了解 OpenStack 是如何来为虚拟机提供存储空间的。

相关知识

1. 基本概念

在 OpenStack Folsom 版本之后，OpenStack 通过 Cinder 服务为云平台提供块逻辑卷服务。逻辑卷通过 RAID 与 LVM 等方法，对数据提供了保护功能，并且使用并行写入和 SAN 架构大幅提升了读写速率。另外，通过将若干磁盘进行组合，形成一个大容量的逻辑卷并对外提供服务，极大地提高了容量。

Cinder 是 OpenStack Block Storage（OpenStack 块存储）的项目名称，Cinder 的核心功能是对卷的管理，允许对卷、卷的类型、快照进行处理。然而，它并没有实现对块设备的管理和实际服务（提供逻辑卷），而是通过后端的统一存储接口支持不同块设备厂商的块存储服务，实现其驱动支持并于 OpenStack 进行整合。Cinder 可以支持如 NetAPP、SolidFire、华为、EMC 和 IBM 等知名存储厂商以及众多开源块存储系统，如图 7-1 所示。

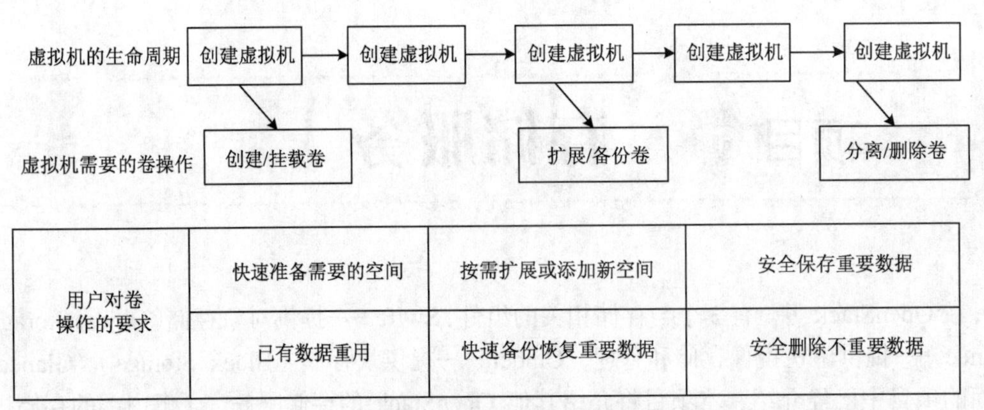

图 7-1 Cinder 项目示意图

对于本地存储，Cinder 通过 Cinder Volume 子服务来使用 LVM 驱动。Cinder Volume 通过“LVM”命令创建一个名为“cinder-volumes”的卷组（VG），当该节点接收到创建卷请求的时候，Cinder Volume 在该 VG 上创建 LV，并且用 iSCSI 服务将这个卷作为一个 iSCSI 的存储设备提供给虚拟机。

2. 架构讲解

Cinder 的架构图如图 7-2 所示。

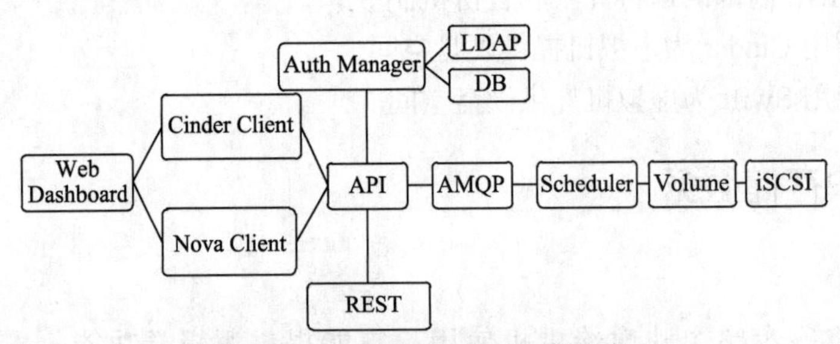

图 7-2 OpenStack Cinder 的架构

Cinder 得到请求后会自动访问块存储服务，它有两个显著的特点，第一，用户必须提出请求，服务才会进行响应，第二，用户可以使用自定义的方式实现半自动化服务。简而言之，Cinder 虚拟化块存储设备池，提供给用户自助服务的 API 用以请求和使用存储池中的资源，而 Cinder 本身并不能获取具体的存储形式或物理设备信息。

① Auth Manage：负责 Cinder 的授权工作，在 OpenStack 中，使用 Keystone 服务来进行授权管理。

② AMQP：高级消息队列协议，用于存储和传递 REST 请求。

③ iSCSI：基于网络的小型计算机系统接口。

④ REST：表征状态转移，它定义了一种软件架构原则，是一种针对网络应用的设计和开发方式，可以降低开发的成本和复杂度，提供系统的可伸缩性。

⑤ Cinder Client：Cinder 的客户端软件，通过 CLI 或 API 等形式访问并使用 Cinder 管理块存储。

3. 配置文件讲解

为了实现 Cinder 通过消息队列存储 REST 请求实现组件间通信，我们需要指定消息队列的类型及存储消息队列的主机。

编辑/etc/cinder/cinder.conf 文件。

```
# vi /etc/cinder/cinder.conf
[DEFAULT]
auth_strategy = keystone
rpc_backend = cinder.openstack.common.rpc.impl_qpid
qpid_hostname = controller
```

Cinder 服务的数据存储在 MySQL 数据库中，需要在 Cinder 中配置连接参数以连接 MySQL 数据库。

```
# mysql -u root -p
[database]
connection = mysql://cinder:000000@controller/cinder
```

```
mysql> CREATE DATABASE cinder;
mysql> GRANT ALL PRIVILEGES ON cinder.* TO 'cinder'@'localhost' IDENTIFIED BY
'CINDER_DBPASS';
mysql> GRANT ALL PRIVILEGES ON cinder.* TO 'cinder'@'%' IDENTIFIED BY '
CINDER_DBPASS ';
# su -s /bin/sh -c 'cinder-manage db sync' cinder
```

为了实现 Cinder 组件的认证管理，需要通过 Keystone 创建 Cinder 用户并设置密码以及权限，并为其创建 API，以便与其他组件进行通信。需要注意的是，Keystone 的 API 有 v1 和 v2 两个版本，为了方便使用，建议分别创建 v1 和 v2 版本的 API。

```
# keystone user-create --name=cinder --pass=CINDER_PASS --email=CINDER_ EMAIL
```

执行结果如下。

Property	Value
email	
enabled	True
id	d49e7607efa24798bc479db4bb90e792
name	cinder
username	cinder

```
# keystone user-role-add --user=cinder --tenant=service --role=admin
# keystone service-create --name=cinder --type=volume --description= "OpenStack
Block Storage"
```

执行结果如下。

Property	Value
description	OpenStack Block Storage

enabled	True
id	b7e27df4df1a4f20a467c027fd1fb6cd
name	cinder
type	volume

```
# keystone endpoint-create  --service-id=$(keystone service-list | awk '/ volume
/ {print $2}')
--publicurl=http://controller:8776/v1/%\(tenant_id\)s
--internalurl=http://controller:8776/v1/%\(tenant_id\)s
--adminurl=http://controller:8776/v1/%\(tenant_id\)s
```

为了实现 Cinder 与 OpenStack 组件间的正常通信，Cinder 需要通过认证管理，对 Cinder 进行授权。Cinder 默认使用本地存储作为后端存储，由于 Cinder 本地存储方式和 iSCSI Target 组件有依赖关系，这里需要配置对应的 iSCSI 组件。

```
# vi /etc/cinder/cinder.conf
[DEFAULT]
auth_strategy = keystone
rpc_backend = cinder.openstack.common.rpc.impl_qpid
qpid_hostname = controller
control_exchange = cinder
notification_driver = cinder.openstack.common.notifier.rpc_notifier
iscsi_helper = tgtadm
[keystone_authtoken]
auth_uri = http://controller:5000
auth_host = controller
auth_protocol = http
auth_port = 35357
admin_user = cinder
admin_tenant_name = service
admin_password = 000000
```

4. LVM 技术

LVM（Logical Volume Manager）是逻辑卷管理的简称，它是 Linux 系统环境下对磁盘分区进行管理的一种机制。现在不仅仅是 Linux 系统可以使用 LVM 这种磁盘管理机制，其他的类 UNIX 操作系统，以及 Windows 操作系统都有类似于 LVM 的磁盘管理软件。

LVM 的工作原理其实很简单，就是将底层的物理硬盘抽象地封装起来，然后以逻辑卷的方式呈现给上层应用。在传统的磁盘管理机制中，上层应用是直接访问文件系统，从而对底层的物理硬盘进行读取的，而在 LVM 中，其通过对底层的硬盘进行封装，当对底层的物理硬盘进行操作时，不再是针对于分区进行操作，而是通过逻辑卷来对其进行底层的磁盘管理操作。例如，增加一个物理硬盘，这个时候上层的服务是感觉不到的，因为呈现给上层服务的是以逻辑卷的方式。LVM 最大的特点就是可以对磁盘进行动态管理。逻辑卷的大小是可以动态调整的，而且不会丢失现有的数据，如果用户新增加了硬盘，也不会改变现有上层的逻辑卷。作为一个动态磁盘管理机制，逻辑卷技术大大提高了磁盘管理的灵活性，如图 7-3 所示。

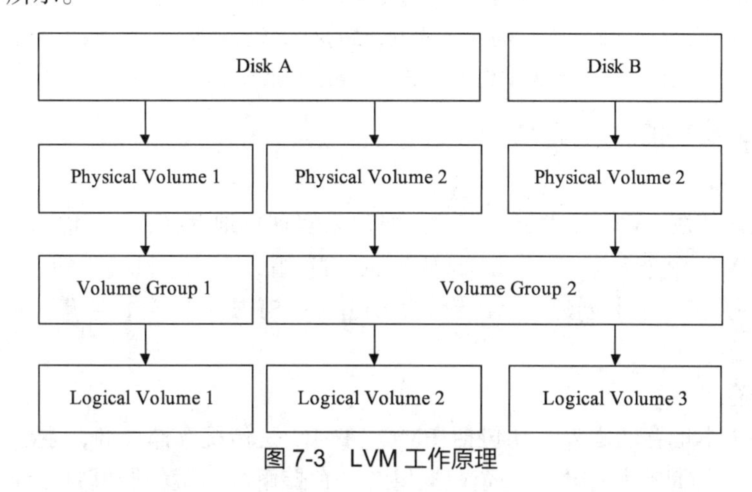

图 7-3 LVM 工作原理

5. iSCSI 技术

iSCSI（Internet SCSI）技术由 IBM 公司研发，是一个供硬件设备使用，可以在网络协议（主要指 TCP/IP 协议栈）的上层运行的 SCSI 指令集。这种指令集可以实现在 IP 网络上运行 SCSI 协议，使其能够在诸如高速吉比特以太网上进行路由选择。SCSI 协议主要是在主机和存储设备之间传送命令、状态和块数据，可以说是当今存储技术最重要的协议之一。iSCSI 技术是一种新型存储技术，该技术允许将现有的 SCSI 接口（支持 SCSI 协议的小型计算机系统接口）与以太网（Ethernet）技术结合，使服务器可以使用 IP 网络的存储装置互相交换资料。

iSCSI 基于 TCP/IP 协议栈，用来建立和管理网络存储设备，以及服务器和客户之间的相互通信，并创建存储网络（SAN）。SAN 可以让 SCSI 协议应用在高速数据传输的网络中，这种传输以数据块级别在多个存储网络之间进行。iSCSI 是基于 C/S 架构的，它的主要功能是在 TCP/IP 网络上的主机系统（启动器 Initiator）和存储设备（目标器 Target）之间进行大量的数据封装和可靠传输，如图 7-4 所示。

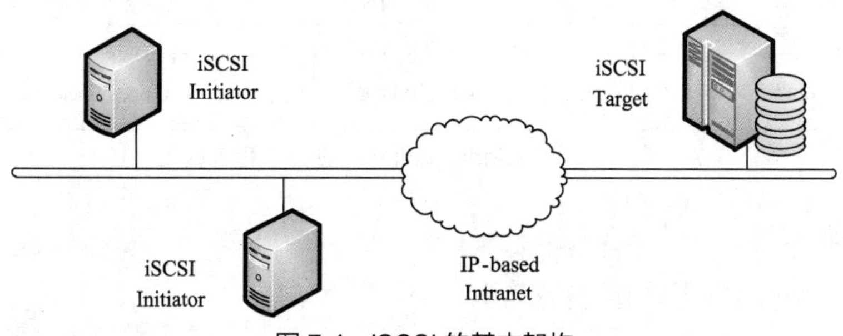

图 7-4 iSCSI 的基本架构

6. Cinder 基本服务

① API Service：负责接受和处理 REST 请求，并将请求放入消息队列中。

② Scheduler Service：负责处理队列中的任务，并根据预先制定的调度策略（优先活跃节点）选择合适的存储节点来执行任务。

③ Volume Service：该服务运行在各个存储节点之上，管理存储空间，每一个存储节点都有一个 Volume Service，从而构建一个庞大的存储资源池。Volume Service 本身不实现存储功能，而由 Cinder 存储后端（Backend Drivers）来实现。

7. Cinder 支持的后端存储类型

（1）本地存储

Cinder 默认使用 LVM 技术配合 iSCSI 协议来实现本地存储，LVM 驱动需要在云主机中事先用“LVM”命令创建一个 Cinder Volume 的卷组，当该主机接收到创建卷的请求时，Cinder Volume 会在卷组上创建一个逻辑卷，并用 iSCSI 技术将这个卷组作为一个 iSCSI 的 Target 提供给云主机。

（2）其他存储

目前，除了本地存储之外，还包括 EMC（EMC 是传统存储厂商，主要面对企业级用户）、Netapp（目前是为数据密集型企业提供统一存储解决方案的居世界最前列的公司）和华为存储。

（3）Cinder Volume 创建流程

当用户从 Cinder Client 发送一个创建卷组请求之后，会将创建的类型、大小和种类等信息通过 RESTFUL 接口来访问 Cinder API，API 接受请求之后就会把 Client 传送的请求进行解析，之后通过 RPC 将请求发送给 Cinder Scheduler，以选择合适的 Volume 节点，节点选择完毕后，再次通过 RPC 发送给 Volume，之后调用 Driver 来创建 Client 要求的卷组，如果 Cinder 需要 Backup，这时候就需要调用 RPC 进行 Backup 操作，如图 7-5 所示。

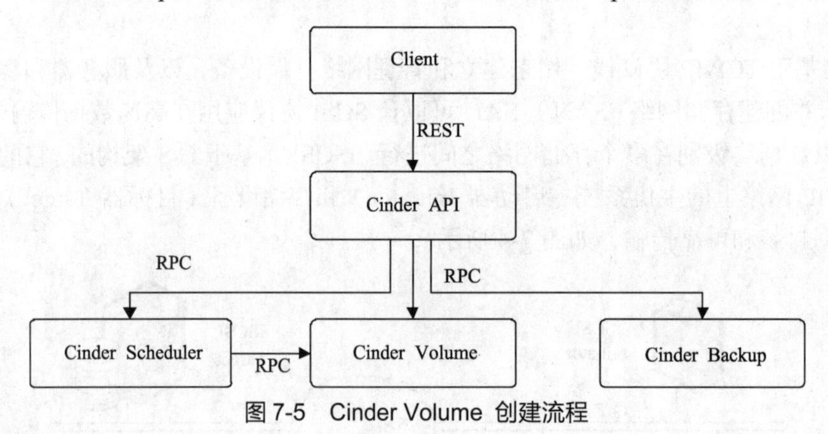

图 7-5　Cinder Volume 创建流程

任务实现

1. 对 Cinder 后端逻辑卷进行扩容

Cinder 后端默认使用 LVM 逻辑卷进行存储，当卷资源不够时，会导致创建 Cinder 卷失败，此时需要对 LVM 逻辑卷进行扩容，从而解决 Cinder Volume 空间不足的问题。

（1）创建一个 100GB 的云硬盘 extend-demo

```
# cinder create --display-name cinder-volume-demo 100
```

执行结果如下。

Property	Value
attachments	[]
availability_zone	Nova
bootable	False
created_at	2016-08-11T02:03:47.063888
display_description	None
display_name	cinder-volume-demo
encrypted	False
id	9b675bcd-2c8a-4786-9bf2-e40b0afa01ef
metadata	{}
size	100
snapshot_id	None
source_volid	None
status	creating
volume_type	None

（2）通过“cinder-list”命令查看云硬盘信息

```
# cinder list
```

执行结果如下。

ID	Status	Display Name	Size	Volume Type
793bc8e6-62e6-463a-bl5d-24a7a13cf420	Available	Test	2	None
9b675bcd-2c8a-4786-9bf2-e40b0afa01ef	error	cinder-volume-demo	100	None

该云硬盘创建失败，状态为 error。用户可以在存储节点运行“vgdisplay”命令查看逻辑卷空间。

```
# vgdisplay
```

执行结果如下。

```
VG Name               cinder-volumes
System ID
Format                lvm2
Metadata Areas        1
Metadata Sequence No  5
VG Access             read/write
VG Status             resizable
```

```
MAX LV               0
Cur LV               3
Open LV              0
MAX LV               0
Cur LV               1
Act PV               1
VG Size              48.82 GiB
PE Size              4.00 MiB
Total PE             12499
Alloc PE / Size      3584 / 14.00 GiB
Free PE / Size       8915 / 34.82 GiB
VG UUID              60iL1Z-X1iF-6fz7-B9k1-CYBe-zBZC-MwhRY3
```

通过查看命令发现，VG 的总容量为 48.48GB，另外还发现创建了一张 1GB 的存储卷以及 PE 占用了 4MB。我们的空间已经不足 47GB，所以无法再创建新的云硬盘，因此需要对 VG 进行扩容。

（3）对 LVM 卷进行扩容

① 通过“pvcreate”命令创建 PV。

```
# pvcreate /dev/md126p2
Physical volume "/dev/md126p2" successfully created
```

② 通过“vgextend”命令扩展已有 VG 的容量。

```
# vgextend cinder-volumes /dev/md126p2
Volume group "cinder-volumes" successfully extended
```

③ 查看扩展后的 VG 容量。

```
# vgdisplay
VG Name              cinder-volumes
System ID
Format                    lvm2
Metadata Areas            2
Metadata Sequence No      6
VG Access                 read/write
VG Status                 resizable
MAX LV                    0
Cur LV                    3
Open LV                   0
MAX LV                    0
Cur LV                    2
Act PV                    2
VG Size                   166.01GiB
PE Size                   4.00 MiB
```

```
Total PE                42498
Alloc PE / Size         3584 / 14.00 GiB
Free PE / Size          38914 / 152.01 GiB
VG UUID                 60iL1Z-X1iF-6fz7-B9k1-CYBe-zBZC-MwhRY3
```

④ 重启 Cinder Volume 服务后再次创建该云硬盘。

```
# cinder create --display-name cinder-volume-demo 100
```

执行结果如下。

Property	Value
attachments	[]
availability_zone	nova
bootable	false
created_at	2016-08-11T02:03:47.063888
display_description	None
display_name	cinder-volume-demo
encrypted	false
id	0643885c-1dc7-4f9f-9309-382c54cc6790
metadata	{}
size	100
snapshot_id	None
source_volid	None
status	creating
volume_type	None

```
# cinder list
```

执行结果如下，“cinder-volume-demo”创建成功。

ID	Status	Display Name	Size
0643885c-1dc7-4f9f-9309-382c54cc6790	available	cinder-volume-demo	100
793bc8e6-62e6-463a-b15d-24a7a13cf420	available	test	2

2. 指定 Cinder 卷类型

上文已经提到，Cinder 中的卷类型可以用于标识卷，可以通过命令行的方式创建、删除或查看卷类型。

（1）创建 type 标识的卷类型

可以通过“cinder type-create”命令来创建卷类型，创建一个名为“lvm”的卷类型。

```
# cinder type-create lvm
```

执行结果如下。

ID	Name
9cca0efb-3518-456c-9202-65df48fc5171	lvm

（2）查询 type 标识的卷类型

可以通过“cinder type-list”命令来查看现有的卷类型。

```
# cinder type-list
```

执行结果如下。

ID	Name
9cca0efb-3518-456c-9202-65df48fc5171	lvm

（3）创建并查询 extra_spec 表示的卷类型

除了可以通过 Cinder 创建 type 类型的卷标识外，OpenStack 还允许通过 Cinder 创建 extra_spec 类型的卷类型，两个功能基本类似，extra_spec 利用一组键值对来对 Cinder 卷进行标识。

可以通过“cinder type-key lvm set volume_backend_name=LVM_iSCSI”命令来创建键为“volume_backend_name”，值为“LVM_iSCSI”的 extra_spec 类型标识。

通过“cinder extra-specs-list”命令来查看已创建的 extra_spec 标识。

```
# cinder extra-specs-list
```

执行结果如下。

ID	Name	extra_specs
9cca0efb-3518-456c-9202-65df48fc5171	lvm	{}

（4）创建云硬盘

下面以 type 标识为例，创建一块带“lvm”标识的云硬盘，命令如下。

```
# cinder create --display-name type_test_demo --volume_type lvm 1
```

执行结果如下。

Property	Value
attachments	[]
availability_zone	nova
bootable	false
created_at	2016-08-11T02:03:47.063888
display_description	None
display_name	type_test_demo
encrypted	False
id	88952cca-11b4-48de-bb3f-d8fb7ba84bbe
metadata	{}
size	1
snapshot_id	None
source_volid	None
status	creating
volume_type	lvm

创建成功后可以通过命令查看结果，可以看到该卷的 volume_type 字段已修改为“lvm”。查询命令如下。

```
# cinder show type_test_demo
```

查询结果如下。

Property	Value
attachments	[]
availability_zone	nova
bootable	false
created_at	2016-08-11T02:03:47.063888
display_description	None
display_name	type_test_demo
encrypted	False
id	88952cca-11b4-48de-bb3f-d8fb7ba84bbe
metadata	{}
os-vol-host-attr:host	compute
os-vol-mig-status-attr:migstat	None
os-vol-mig-status-attr:name_id	None
os-vol-tenant-attr:tenant_id	a3c430debe7e48d4a2f58c96dd89746b
size	1
snapshot_id	None
source_volid	None
status	available
volume_type	lvm

3. Cinder 的 CLI 命令行使用

Cinder 作为 OpenStack 平台的块存储组件，提供了一系列存储操作的 CLI 命令行用于管理存储卷，可以灵活地对存储卷进行创建、扩容、删除和加密等操作。

（1）创建 Cinder 存储卷

用户可以通过“CLI”命令创建一个简单的 Cinder 存储卷，命令如下。

```
# cinder create --display-name cinder-volume-demo 1
```

输出结果如下所示。

Property	Value
attachments	[]
availability_zone	nova
bootable	false
created_at	2016-08-11T02:03:47.063888
display_description	None
display_name	Cinder-volume-demo
encrypted	False

id	0643885c-1dc7-4f9f-9309-382c54cc6790
metadata	{}
size	1
snapshot_id	None
source_volid	None
status	creating
volume_type	None

该命令用法如下。

```
cinder create [--snapshot-id <snapshot-id>]
            [--source-volid <source-volid>]
[--image-id <image-id>]
            [--display-name <display-name>]
            [--display-description <display-description>]
            [--volume-type <volume-type>]
            [--availability-zone <availability-zone>]
            [--metadata [<key=value> [<key=value> ...]]]
            <size>
Positional arguments:
 <size>              Size of volume in GB
```

用户可以通过--display 参数指定卷名，然后指定卷组大小，创建一个最简单的云存储卷。

（2）应用上述命令创建特定的镜像，参数说明如下

〈--image-id〉参数允许用户从一个已创建的镜像来构建一个带镜像内容的 Cinder 卷，一般可以用于启动虚拟机操作系统。

〈--snapshot-id〉参数允许用户从特定的快照来创建Cinder卷，具体用法和--image-id类似。

[--source-volid]参数允许用户从已创建的 Cinder 卷来创建一张内容相同的卷，可以用于备份或启动系统。

[--display-description]参数允许用户指定镜像的描述信息。

[--volume-type]参数可以指定卷类型，卷类型是卷的一个标识，各个 OpenStack 发行者可以根据自身对系统的约束来定义卷类型的使用。在扩展延伸部分会对 Cinder 卷类型进行详细阐述。

[--availability-zone]参数可以为创建的 Cinder 卷指定可用域，不同可用域的资源会被隔离，互不干扰。

[--metadata]参数可以允许为创建的 Cinder 卷指定属性信息。

（3）删除指定的 Cinder 卷

删除 Cinder 卷的方法比较简单，用户可以通过“cinder delete <volume> [<volume> ...]”命令来删除一个或多个 Cinder 卷。

（4）创建 Cinder 卷快照

用户可以通过如下命令来创建一个简单的 Cinder 卷快照。

```
# cinder snapshot-create --display-name snapshot_demo cinder-volume-demo
```

Property	Value
created_at	2016-08-11T02:03:47.063888
display_description	None
display_name	snapshot -demo
id	ec29d31e-996b-4b5f-9ee7-3bd337a5d89a
metadata	{}
size	1
status	creating
volume_id	0643885c-1dc7-4f9f-9309-382c54cc6790

该命令用法如下。

```
cinder snapshot-create [--force <True|False>]
                       [--display-name <display-name>]
                       [--display-description <display-description>]
                                 <volume>
Optional arguments:
                            --force <True|False>
          --display-name <display-name>
          --display-description <display-description>
```

其中，--force 参数用于强制创建该快照，无论该卷是否被虚拟机所使用。

（5）查看 Cinder 卷、快照信息

可以通过“cinder-list”命令查看镜像列表。

```
# cinder list
```

输出结果如下。

ID	Status	Display Name	Size	Volume Type
0643885c-1dc7-4f9f-9309-382c54cc6790	available	Cinder-volume-demo	1	None
793bc0e6-62e6-463a-b15d-24a7a13cf420	available	test	2	None
88952cca-11b4-48de-bb3f-d8fb7ba84bbe	available	type_test_demo	1	lvm

还可以通过“cinder show <volume>”命令来查看指定云硬盘的详细信息，命令如下。

```
# cinder show 0643885c-1dc7-4f9f-9309-382c54cc6790
```

输出结果如下。

Property	Value
attachments	[]
availability_zone	nova
bootable	false
created_at	2016-08-11T02:03:47.063888
display_description	None
display_name	Cinder-volume-demo
encrypted	False

id	0643885c-1dc7-4f9f-9309-382c54cc6790
metadata	{}
os-vol-host-attr:host	compute
os-vol-mig-status-attr:migstat	None
os-vol-mig-status-attr:name_id	None
os-vol-tenant-attr:tenant_id	a3c430debe7e48d4a2f58c96dd89746b
size	1
snapshot_id	None
source_volid	None
status	available
volume_type	None

（6）将已创建的 Cinder 卷挂载到虚拟机实例

云硬盘创建完成后，可以将它挂载到云主机实例上。首先选择右侧标签栏中的“云硬盘”标签，单击“更多”按钮，选择“编辑挂载”选项，如图 7-6 所示。

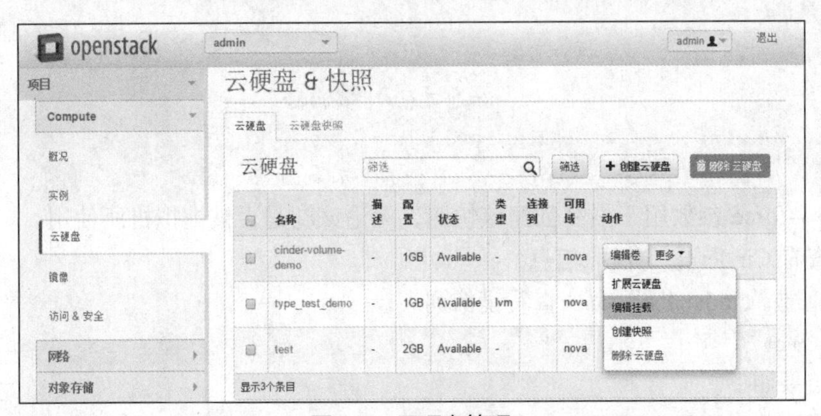

图 7-6　云硬盘管理

在“连接到云主机”下拉框中选择创建的云主机，单击“连接云硬盘”按钮，如图 7-7 所示。

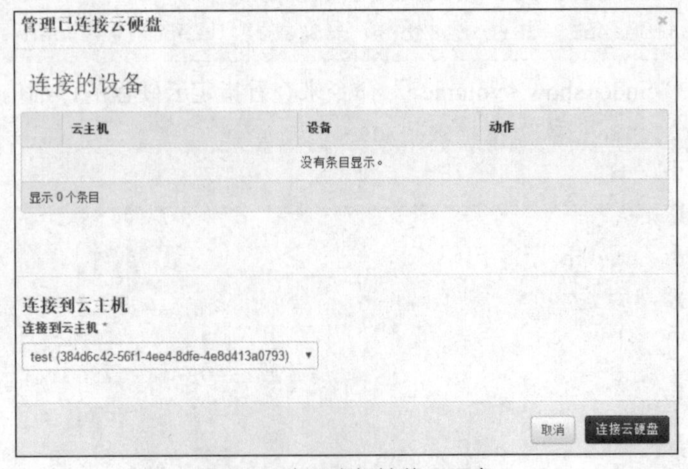

图 7-7　为云主机挂载云硬盘

连接完成后，可以看到该云硬盘的状态为“In-Use”，即表示挂载成功，如图 7-8 所示。

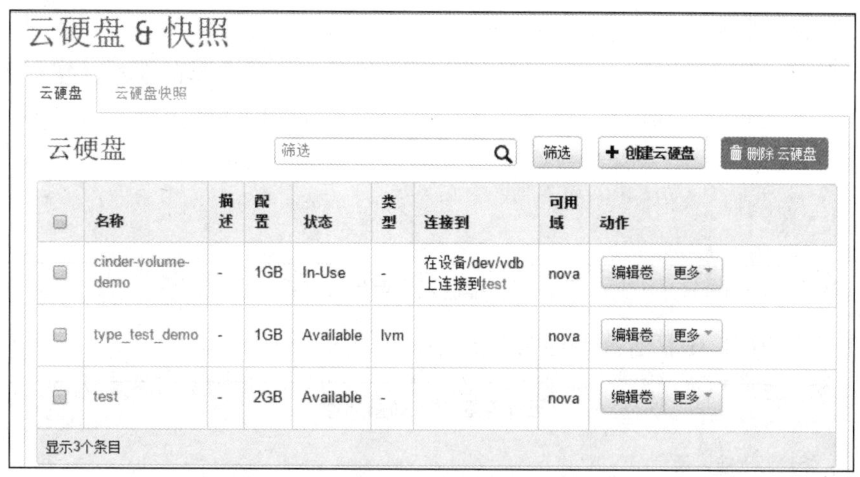

图 7-8　挂载成功

4. Dashboard 完成块存储任务

（1）创建名为“Volume_test”，大小为 1GB 的云硬盘，并在 Cloud_test 实例上挂载

① 进入 Dashboard 界面找到“项目”选项。

② 打开 compute 面板，选择云硬盘。

③ 选择“创建云硬盘”。

④ 在弹出的窗口中输入“云硬盘名称”“描述”“类型”“云硬盘源自”和“可用域”等信息，如图 7-9 所示。

图 7-9　创建云硬盘

⑤ 在动作栏目中，单击“更多”按钮，选择“编辑挂载”选项，如图 7-10 所示。

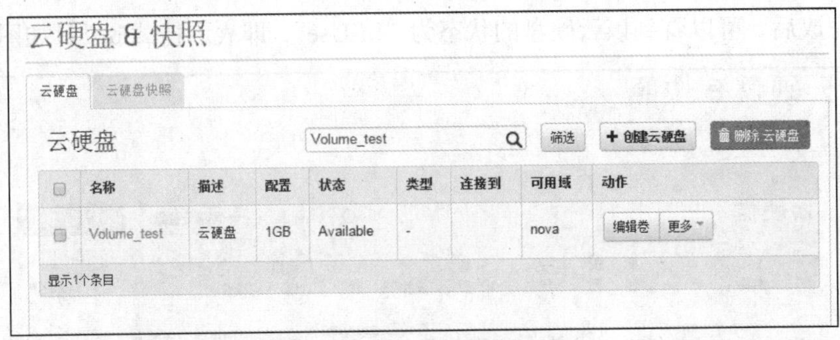

图 7-10　挂载云硬盘到云主机

⑥ 在弹出的窗口中，选择要连接的云主机，如图 7-11 所示。

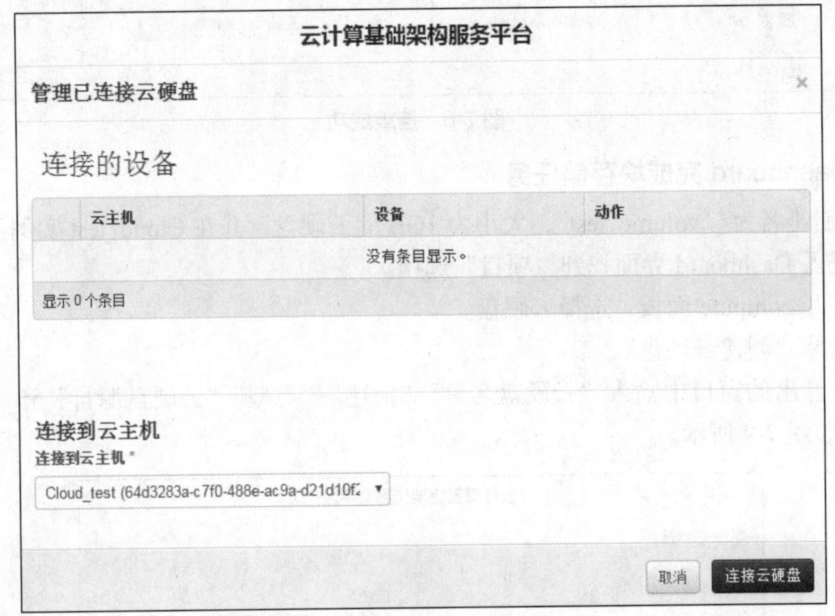

图 7-11　选择要连接的云主机

⑦ 连接完成，可以看到云硬盘处于“In-Use”状态，如图 7-12 所示。

图 7-12　查看连接成功的云硬盘

⑧ 在实例中打开相应云主机的控制台，查看设备是否存在，如图 7-13 所示。

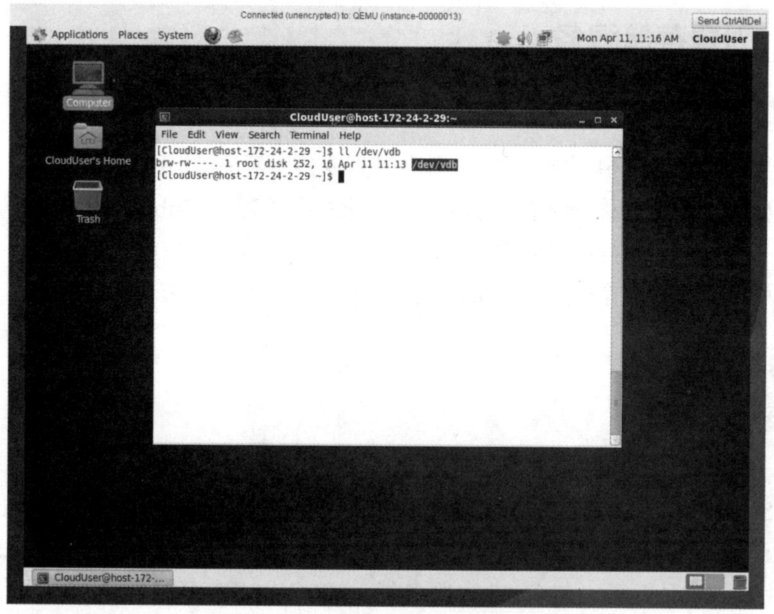

图 7-13　在实例中查看设备信息

（2）保障实例的安全性，创建 Cinder 的快照资源，做到实时恢复和备份

① 找到对应的云硬盘，单击“更多”按钮，选择“创建云硬盘快照”（在为云硬盘创建快照之前，最好先断开与虚拟机的连接），如图 7-14 所示。

② 在弹出的窗口中，设置快照名称。

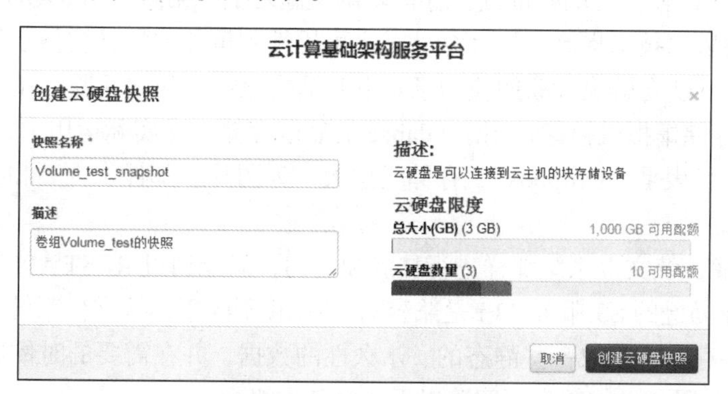

图 7-14　创建云硬盘快照

③ 进入“云硬盘快照”选项卡中查看快照资源，如图 7-15 所示。

图 7-15　查看快照资源

任务二 对象存储服务

任务要求

现企业内部拥有 3 个用户组，3 个用户组需要内部存储服务器作为内部存储资源，需要每个租户创建以自己名称命名的存储容器，创建后缀为“_Public”的对外服务的存储，创建后缀为“_Private”的自己使用的容器。

为保证内部数据的安全性，将 Glance、Cinder 的后端存储修改为 Swift。

相关知识

1. 发展现状

在国外，成功的商用案例有 Rackspace 公司通过结合 Swift 和 Nova 提供的 IaaS 云服务，微软公司的 SharePoint 后端支持，韩国电信公司（KT）推出的云服务等。

在国内尚未有商用的成功案例，大部分都处于实验阶段，上海交大信息中心用 OpenStack 做了一个私有云，包含 50 台服务器，500GB SSD 存储，100TB 块存储和 400TB 对象存储。

2011 年 9 月 6 日，首届开源云 OpenStack 峰会在上海举行，由此可见，OpenStack 在国内的研究还比较热门。

2. 基本概念

Swift 构筑在比较便宜的标准硬件存储基础设施之上，无需采用 RAID，它通过在软件层面引入一致性散列技术提高数据冗余性、高可用性和可伸缩性，支持多租户模式、容器和对象读写操作，适合解决互联网应用场景下非结构化数据存储问题。在 OpenStack 中，Swift 主要用于存储虚拟机镜像，用于 Glance 的后端存储。在实际运用中，Swift 的典型应用是网盘系统，代表是“Dropbox”，存储类型大多为图片、邮件、视频和存储备份等静态资源。

Swift 不能像传统文件系统那样进行挂载和访问，只能通过 REST API 来访问数据，并且这些 API 与亚马逊的 S3 服务 API 是兼容的。Swift 不同于传统文件系统和实时数据存储系统，它适用于存储、获取一些静态的、永久性的数据，并在需要的时候进行更新。

在了解 Swift 服务之前首先要明确以下 3 个基本概念。

（1）Account

出于访问安全性考虑，使用 Swift 系统时，每个用户必须有一个账号（Account）。只有通过 Swift 验证的账号才能访问 Swift 系统中的数据。提供账号验证的节点被称为 Account Server。Swift 中由 Swauth 提供账号权限认证服务。

用户通过账号验证后将获得一个验证字符串（Authentication Token），后续的每次数据访问操作都需要传递这个字符串。

（2）Container

Swift 中的 Container 可以类比 Windows 操作系统中的文件夹或者类 UNIX 操作系统中的目录，用于组织管理数据，所不同的是，Container 不能嵌套，数据都以 Object 的形式存放在 Container 中。

（3）Object

Object（对象）是 Swift 中的基本存储单元。一个对象包含两部分：数据和元数据（Metadata）。其中，元数据包括对象所属 Container 名称、对象本身名称，以及用户添加的自定义数据属性（必须是 Key-value 格式）。

对象名称在 URL 编码后，要求小于 1024 字节。用户上传的对象最大是 5GB，最小是 0 byte。用户可以通过 Swift 内建的大对象支持技术获取超过 5GB 的大对象。对象的元数据不能超过 90 个 Key-value 对属性，并且这些属性的总大小不能超过 4KB。

Account、Container 和 Object 是 Swift 系统中的 3 个基本概念，三者的层次关系是，一个 Account 可以创建任意多个 Container，一个 Container 中可以包含任意多个 Object。可以简单理解为，一个 Tenant 拥有一个 Account，Account 下存放 Container，Container 下存储 Object。

在 Swift 系统中，集群被划分成多个区（Zone），区可以是一个磁盘、一个服务器、一台机柜甚至一个数据中心，每个区中有若干个节点。Swift 将 Object 存储在节点上，每个节点都是由多个硬盘组成的，并保证对象在多个节点上有备份（默认情况下，Swift 会给所有数据保存 3 个复本）以及这些备份之间的一致性。备份将均匀地分布在集群服务器上，并且系统可保证各个备份分布在不同区的存储设备上，这样可以提高系统的稳定性和数据的安全性。它可以通过增加节点来线性地扩充存储空间，当一个节点出现故障时，Swift 会在其他正常节点对出故障节点的数据进行备份。

3. Swift 服务优势

（1）数据访问灵活性

Swift 通过 REST API 来访问数据，可以通过 API 实现文件的存储和管理，使得资源管理实现自动化。同时，Swift 将数据放置于容器内，我们可以创建公有的容器和私有的容器。自由的访问控制权限既允许用户间共享数据，也可以保存隐私数据。Swift 对所需的硬件没有刻意的要求，可以充分利用商用的硬件节约单位存储的成本。

（2）高数据持久性

Swift 提供多重备份机制，拥有极高的数据可靠性，数据存放在高分布式的 Swift 系统中，几乎不会丢失，Swift 在 5 个 Zone、5 × 10 个存储节点、数据复制 3 份时，数据持久性的 SLA 能够达到 10 个 9，即存储 1 万个文件到 Swift 中，经过 10 万年后，可能会丢失一个文件，这种文件丢失程度几乎可以忽略不计。

（3）极高的可拓展性

Swift 通过独立节点来形成存储系统，它在数据量的存储上做到了无限拓展。另外，Swift 的性能也可以通过增加 Swift 集群来实现线性提升，所以 Swift 很难达到性能瓶颈。

（4）无单点故障

由于 Swift 的节点独立的特点，在实际工作时，不会发生传统存储系统的单点故障，传统系统即使通过 HA 来实现热备，在主节点出现问题时，还是会影响整个存储系统的性能。而在 Swift 系统中，数据的元数据（Metadata）是通过 Ring 算法随机均匀分布的，且元数据也会保存多份，对于整个 Swift 集群而言，没有单点的角色存在。

4．架构解析

Swift 采用了 REST 架构，REST 是 Roy Fielding 博士在 2000 年他的博士论文中提出来的一种软件架构风格。REST 是一种轻量级的 Web Service 架构风格，其实现和操作明显比 SOAP 和 XML-RPC 更为简洁，可以完全通过 HTTP 实现，它还可以利用 Cache 来提高响应速度，性能、效率和易用性上都优于 SOAP。

REST 架构遵循了 CRUD 原则，CRUD 原则对于资源只需要 4 种行为（Create（创建）、Read（读取）、Update（更新）和 Delete（删除）），就可以完成对其的操作和处理。这 4 个操作是一种原子操作，即一种无法再分的操作，通过它们可以构造复杂的操作过程，正如数学上的四则运算是数字的最基本的运算一样。

REST 架构让人们真正理解网络协议 HTTP 的本来面貌，对资源的操作，包括获取、创建、修改和删除，正好对应 HTTP 提供的 GET、POST、PUT 和 DELETE 方法，因此 REST 把 HTTP 对一个 URL 资源的操作限制在 GET、POST、PUT 和 DELETE 这 4 个之内。这种针对网络应用的设计和开发方式，可以降低开发的复杂性，提高系统的可伸缩性。

由于其简洁方便性，越来越多的 Web 服务开始采用 REST 风格设计和实现。例如，Amazon.com 提供接近 REST 风格的 Web 服务进行图书查找。

因为 Swift 采用 REST 架构，所以用户不能像普通的文件系统那样对数据进行访问，必须通过它提供的 API 来访问操作数据，如图 7-16 所示。图 7-17 和图 7-18 分别展示了 Swift 的上传和下载操作。Rackspace 还对这些 API 做了不同语言的封装绑定，以方便开发者进行开发。目前支持的语言有 PHP、Python、Java、C#、.NET 和 Ruby。

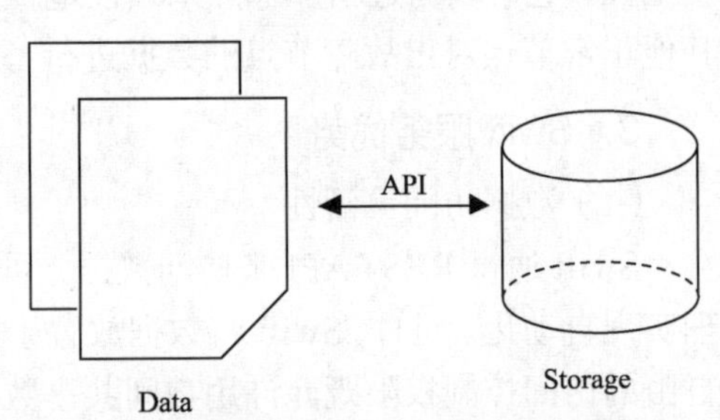

图 7-16　通过 API 访问 Swift 存储的数据

使用 HTTPS（SSL）和对象存储进行交互时，也可以使用标准的 HTTP API 调用来完成操作。同样，也可以使用特定语言的 API，如使用 RESTFUL API。

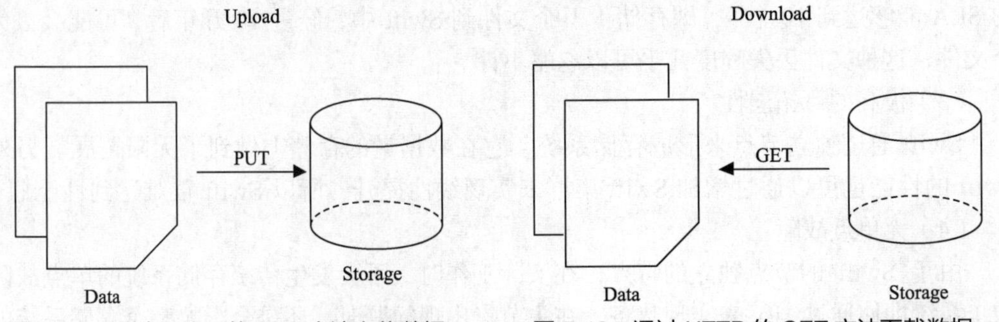

图 7-17　通过 HTTP 的 PUT 方法上传数据　　图 7-18　通过 HTTP 的 GET 方法下载数据

通过 API 可使用 PUT 方法将 Data 数据上传到存储系统中，如果需要下载数据则通过 GET 方法将存储系统中的数据下载下来。

Swift 集群主要包含认证节点、代理节点和存储节点。认证节点主要负责对用户的请求进行授权，只有通过认证节点授权的用户才能操作 Swift 服务。因为 Swift 是 OpenStack 的子项目之一，所以一般用 Keystone 服务作为 Swift 服务的认证服务。代理节点用于和用户交互，接受用户的请求，并且给用户做出响应。Swift 服务所存储的数据一般都放在数据节点上。

图 7-19 是 Swift 架构图，通过图片来详细讲解 Swift 服务的各个组件及其功能。

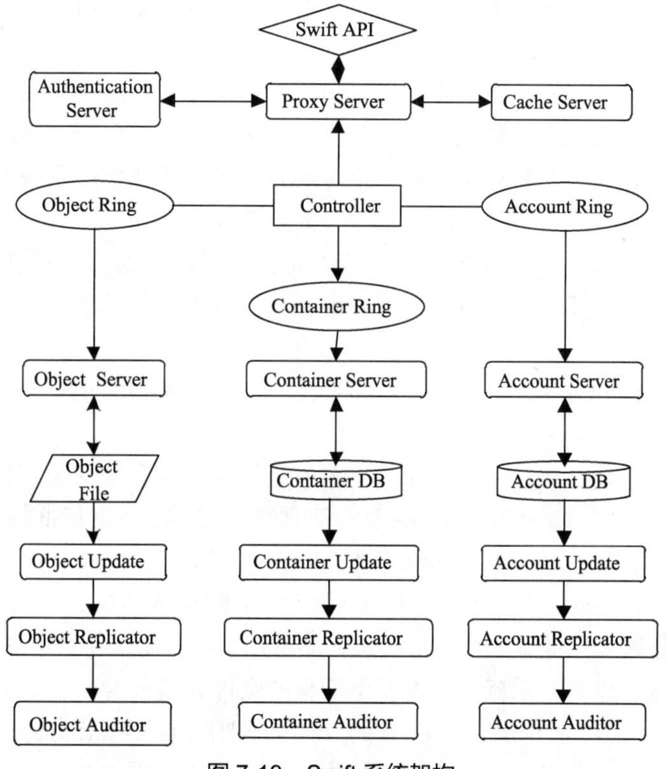

图 7-19 Swift 系统架构

① 代理服务（Proxy Server）：对外提供对象服务 API，会根据环的信息来查找服务地址并转发用户请求至相应的账户、容器或者对象服务。由于采用无状态的 REST 请求协议，可以进行横向扩展来实现负载均衡。

② 认证服务（Authentication Server）：验证访问用户的身份信息，并获得一个对象访问令牌（Token），其在一定的时间内会一直有效；验证访问令牌的有效性并缓存下来直至时间过期。

③ 缓存服务（Cache Server）：缓存的内容包括对象服务令牌、账户和容器的存在信息，但不会缓存对象本身的数据；缓存服务可采用 Memcached 集群，Swift 会使用一致性散列算法来分配缓存地址。

④ 账户服务（Account Server）：提供账户元数据和统计信息，并维护所含容器列表的服务，每个账户的信息被存储在一个 SQLite 数据库中。

⑤ 容器服务（Container Server）：提供容器元数据和统计信息，并维护所含对象列表

的服务，每个容器的信息也存储在一个 SQLite 数据库中。

⑥ 对象服务（Object Server）：提供对象元数据和内容服务，每个对象的内容会以文件的形式存储在文件系统中，元数据会作为文件属性进行存储，建议采用支持扩展属性的 XFS 文件系统。

⑦ 复制服务（Replicator）：检测本地分区副本和远程副本是否一致，具体是通过对比散列文件和高级水印的一致性来完成的，发现不一致时会采用推式（Push）方法更新远程副本，例如，对象复制服务会使用远程文件复制工具 Rsync 来进行同步；另外一个任务是确保被标记删除的对象从文件系统中移除。

⑧ 更新服务（Updater）：当对象由于高负载的原因而无法立即更新时，任务将会被序列化到在本地文件系统中进行排队，以便服务恢复后进行异步更新，例如，成功创建对象后，容器服务器没有及时更新对象列表，这时候容器的更新操作就会进入排队中，更新服务会在系统恢复正常后扫描队列并进行相应的更新处理。

⑨ 审计服务（Auditor）：检查对象、容器和账户的完整性，如果发现比特级的错误，文件将被隔离，并复制其他的副本以覆盖本地损坏的副本，其他类型的错误会被记录到日志中。

⑩ 账户清理服务（Account Reaper）：移除被标记为“删除”的账户，删除其所包含的所有容器和对象。

⑪ 环（Ring）：Ring 是 Swift 最重要的组件，用于记录存储对象与物理位置间的映射关系。在涉及 Account、Container 和 Object 信息时，就需要查询集群的 Ring 信息。Ring 使用 Zone、Device、Partition 和 Replica 来维护这些映射信息。Ring 中每个 Partition 在集群中都（默认）有 3 个 Replica。每个 Partition 的位置由 Ring 来维护，并存储在映射中。Ring 文件在系统初始化时被创建，之后每次增减存储节点时，需要重新平衡一下 Ring 文件中的项目，以保证增减节点时，系统因此而发生迁移的文件数量最少。

⑫ 区域（Zone）：Ring 中引入了 Zone 的概念，即把集群的 Node 分配到每个 Zone 中。其中，同一个 Partition 的 Replica 不能同时放在同一个 Node 上或同一个 Zone 内，以防止所有的 Node 都在一个机架或一个机房中时，一旦发生断电、网络故障等情况，会造成用户无法访问。

5. 一致性散列

海量级别的对象，需要存放在成千上万台服务器和硬盘设备上，这首先要解决的是寻址问题，即如何将对象分布到这些设备地址上。Swift 是基于一致性散列技术的，通过计算可将对象均匀分布到虚拟空间的虚拟节点上，在增加或删除节点时可大大减少需移动的数据量。虚拟空间大小通常采用 2 的 n 次幂，以便于进行高效的移位操作。然后，通过独特的数据结构 Ring（环）再将虚拟节点映射到实际的物理存储设备上，完成寻址过程，如图 7-20 所示。

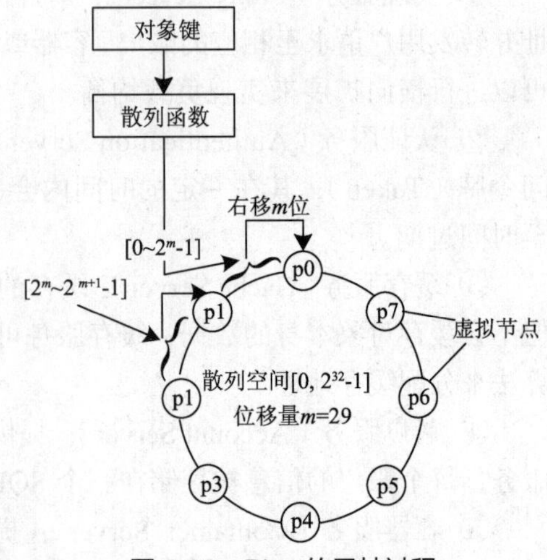

图 7-20　Ring 的寻址过程

如图 7-20 所示，以逆时针方向递增的散列空间有 4 个字节，共 32 位，整数范围是 $0\sim2^{32}-1$，将散列结果右移 m 位，可产生 2^{32-m} 个虚拟节点，例如，$m=29$ 时可产生 8 个虚拟节点。在实际部署的时候需要经过仔细计算得到合适的虚拟节点数，以达到存储空间和工作负载之间的平衡。

6. 数据一致性模型

按照 Eric Brewer 的 CAP（Consistency，Availability，Partition Tolerance）理论，Swift 是无法同时满足 3 个方面的，Swift 通过放弃严格一致性（满足 ACID 事务级别），而最终采用一致性模型（Eventual Consistency），来达到高可用性和无限水平扩展能力。为了实现这一目标，Swift 采用 Quorum 仲裁协议（Quorum 有法定投票人数的含义）：设定 N 为数据的副本总数；W 为写操作被确认接受的副本数量；R 为读操作的副本数量。

强一致性：R+W>N，以保证对副本的读写操作会产生交集，从而保证可以读取到最新版本。如果 W=N，R=1，则需要全部更新，适合大量读少量写操作场景下的强一致性；如果 R=N，W=1，则只更新一个副本，通过读取全部副本来得到最新版本，适合大量写少量读场景下的强一致性。

弱一致性：R+W≤N，如果读写操作的副本集合不产生交集，就可能会读到脏数据。它适合对一致性要求比较低的场景。

Swift 针对的是读写都比较频繁的场景，所以采用了折中的策略，即写操作需要满足至少一半以上成功，即 W>N/2，再保证读操作与写操作的副本集合至少产生一个交集，即 R+W>N。Swift 默认配置是 N=3，W=2>N/2，R=1 或 2，即每个对象会存在 3 个副本，这些副本会尽量被存储在不同区域的节点上，W=2 表示至少需要更新 2 个副本才算写成功；当 R=1 时意味着某一个读操作成功便立刻返回，此种情况下可能会读取到旧版本（弱一致性模型）；当 R=2 时，需要通过在读操作请求头中增加 x-newest=true 参数来同时读取 2 个副本的元数据信息，然后比较时间戳来确定哪个是最新版本（强一致性模型）；如果数据出现了不一致，后台服务进程会在一定时间窗口内通过检测和复制协议来完成数据同步，从而保证数据达到最终一致性，如图 7-21 所示。

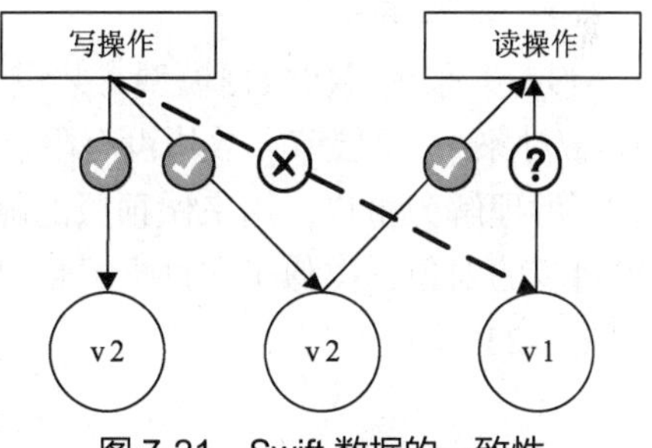

图 7-21　Swift 数据的一致性

7. 环的数据结构

环是为了将虚拟节点（分区）映射到一组物理存储设备上，并提供一定的冗余度而设计的，其数据结构由以下信息组成。

存储设备列表、设备信息包括唯一标识号（ID）、区域号（Zone）、权重（Weight）、IP 地址（IP）、端口（Port）、设备名称（Device）、元数据（Meta）、分区到设备映射表（Replica2part2dev_id 数组）、计算分区号的位移（Part_shift，整数，即图 7-20 中的 m）。

如图 7-22 所示，以查找一个对象的计算过程为例。

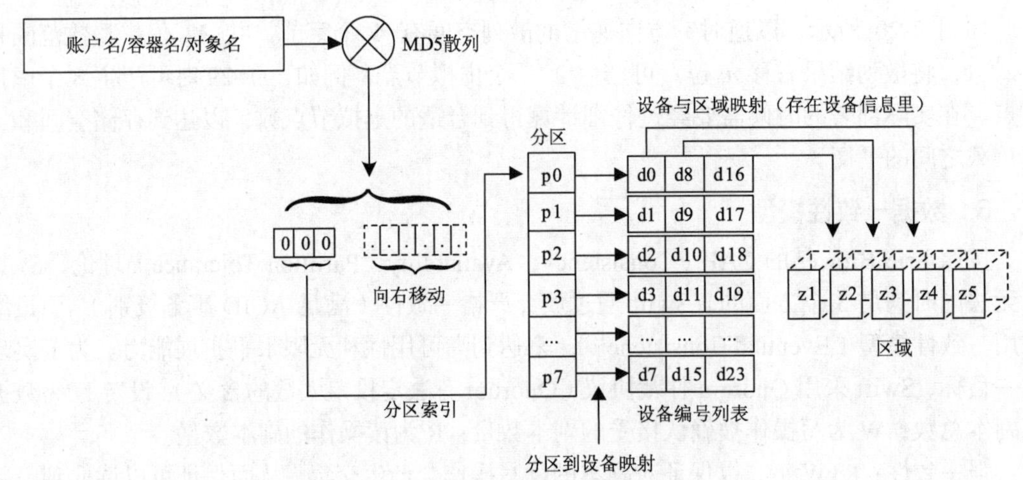

图 7-22　查找一个对象的计算过程

层次结构 account/container/object 作为键，可使用 MD5 散列算法得到一个散列值，对该散列值的前 4 个字节进行右移操作可得到分区索引号，移动位数由上面的 part_shift 设置指定；按照分区索引号在分区到设备映射表（Replica2part2dev_id）里查找该对象所在分区对应的所有设备编号，这些设备会被尽量选择部署在不同区域（Zone）内，区域只是个抽象概念，它可以是某台机器、某个机架，甚至某个建筑内的机群，以提供最高级别的冗余性，建议至少部署 5 个区域；权重参数是个相对值，可以根据磁盘的大小来调节，权重越大表示可分配的空间越多，可部署更多的分区。

Swift 为账户、容器和对象分别定义了的环，查找账户和容器也是同样的过程。

8. 数据模型

Swift 采用层次数据模型，共设 3 层逻辑结构，即 Account/Container/Object（即账户/容器/对象），每层节点数均没有限制，可以任意扩展。这里的账户和个人账户不是一个概念，可理解为租户，用来做顶层的隔离机制，可以被多个个人账户共同使用；容器代表一组封装的对象，类似于文件夹或目录；叶子节点代表对象，由元数据和内容两部分组成，如图 7-23 所示。

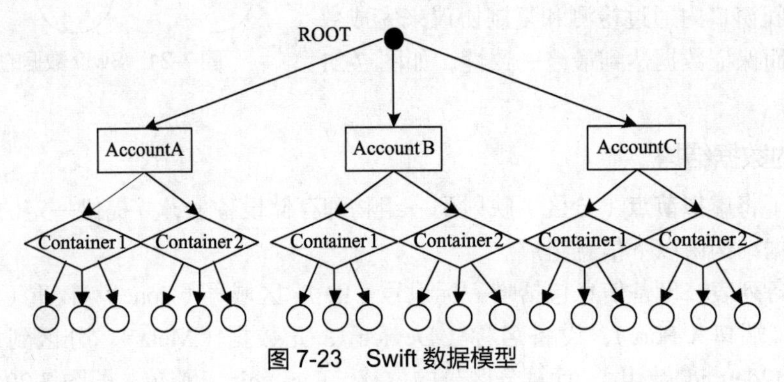

图 7-23　Swift 数据模型

9. 基本命令

Swift 工具是用户与 OpenStack 对象存储（Swift）环境进行通信的命令行接口。它允许一个用户执行多种类型的操作，常用的管理命令有以下几种。

（1）swift stat

功能：根据给定的参数显示账户、对象或容器的信息。

格式：

```
swift stat [container] [object]
```

参数说明如下。

[container]：容器名称。

[object] ：对象名称。

（2）swift list

功能：列出该账户的容器或容器的对象。

格式：

```
swift list [command-options] [container]
```

参数说明如下。

[command-options]：选项。

[container]：容器名称。

（3）swift upload

功能：根据参数将制定的文件或者目录上传到容器内。

格式：

```
swift upload [command-options] container file_or_directory [file_or_ directory]
[...]
```

参数说明如下。

[command-options]：选项。

[container]：容器名称，或者是容器内的目录。

[file_or_directory] ：本地文件系统内的目录或者文件，可同时上传多个目录或文件。

（4）swift post

功能：根据给定的参数升级 Account、Container 或者 Object 的元数据信息。

格式：

```
swift post [command-options] [container] [object]
```

参数说明如下。

[command-options]：选项。

[container]：容器名称。

[object]：对象名称。

（5）swift download

功能：根据给定的参数下载容器中的对象。

格式：

```
swift download [command-options] [container] [object] [object] [...]
```

参数说明如下。

[command-options]：选项。

[container]：容器名称。

[object]：对象名称(可同时下载多个对象)。

（6）swift delete

功能：根据给定的参数删除容器中的对象。

格式：

```
swift delete [command-options] [container] [object] [object] [...]
```

参数说明如下。

[command-options]：选项。

[container]：容器名称。

[object]：对象名称(可同时下载多个对象)。

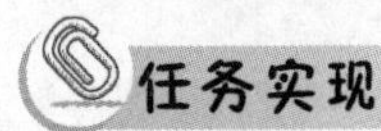

1. 熟悉 Swift 基本操作

通过命令行实现对 Swift 上数据的操作，首先需要创建一个名称为“xiandian”的容器，命令如下。

```
# swift post xiandian
```

有了容器之后，可以查看“xiandian”容器里面的内容，命令如下。

```
# swift list xiandian
```

通过显示结果可以看出，目前“xiandian”容器里面的内容是空的，这时用户希望将本地的“test”目录内容递归上传到“xiandian”容器内，首先要创建“test”目录，并同时新建 3 个文件“iaas.txt”“paas.doc”和“saas.png”，具体命令如下。

```
# mkdir test
# touch iaas.txt
# touch paas.doc
# touch saas.png
```

上传时首先需要上传一个空白的 test 目录，命令如下。

```
# swift upload xiandain test/
```

接下来，可以将“iaas.txt”文件上传到“xiandian”容器内 test 目录内，命令和执行结果如下。

```
# swift upload xiandian/test iaas.txt
[root@controller mnt] # swift upload xiandian/test iaas.txt
iaas.txt
```

换一种方式将剩下的“paas.doc”和“saas.png”递归上传到“xiandian”容器下的 test 目录内，命令和执行结果如下。

```
# mv paas.doc saas.png test/
# swift upload xiandian test/
test/paas.doc
test/saas.png
```

数据在 Swift 集群内保存，随时供用户下载使用，下载“saas.png”文件的命令和执行结果如下。

```
# swift download xiandian/test saas.png
saas.png [auth 0.259s, headers 0.325s, total 0.325s, 0.000 MB/s]
[root@controller mnt]# ll
total 4
-rw-r--r--. 1 root root    0 Aug 12 01:59 saas.png
Drwxr-xr-x. 2 root root 4096 Aug 12 02:04 test
```

目前，磁盘容量有限，需要删除一些相对价值低的数据，以空出更多的空间。由于这时已经将“saas.png”下载到本地，所以暂时将“saas.png”从对象存储服务器中删除，命令和执行结果如下。

```
# swift delete xiandian test/saas.png
[root@controller mnt] # swift delete xiandian test/saas.png
test/saas.png
```

用户还可以通过“swift stat”命令来查看整个Account账户下的Swift状态，命令和执行结果如下。

```
# swift stat
[root@controller ~]# swift stat
Account: AUTH_a3c430debe7e48d4a2f58c96dd89746b
Containers: 2
Objects: 5
Bytes: 0
Accept-Ranges: bytes
X-Timestamp: 1470884707.65905
X-Trans-Id: tx2b3cb1d7db284666bbc96-0057ad68b5
Content-Type: text/plain; charset=utf-8
```

从返回信息中可以看出，账号拥有2个容器、5个对象，还可看出总共占用磁盘大小等信息。接下来，还可以查看容器的具体运行状态。以查看“xiandian”容器为例，命令和执行结果如下。

```
# swift stat xiandian
Account: AUTH_a3c430debe7e48d4a2f58c96dd89746b
Container:xiandian
Objects:3
Bytes:0
Read ACL:
Write ACL:
Sync To:
Sync Key:
Accept-Ranges:bytes
X-Timestamp: 1470884707.72428
```

```
X-Trans-Id: tx6504e55a5dad4e17b31ca-0057ad68ec
Content-Type: text/plain; charset=utf-8
```

从返回的信息中可以看出，目前查看的容器名称为“xiandian”，该容器包含了 3 个对象，还可看出占用磁盘大小等信息。

最后可以查看“xiandian”容器内某个具体对象“test”的状态，该对象是个目录，命令和执行结果如下。

```
$ swift stat xiandian test
Account: AUTH_a3c430debe7e48d4a2f58c96dd89746b
Container: xiandian
Object: test
Content Type: application/octet-stream
Content Length: 0
Last Modified: Fri, 12 Aug 2016 06:03:09 GMT
ETag: d41d8cd98f00b204e9800998ecf8427e
Accept-Ranges: bytes
X-Timestamp: 1470981788.51561
X-Trans-Id: txbbaf6091812b47e7ae217-0057ad692a
```

从返回的信息可以看出，当前对象所在的容器为“xiandian”，当前查看的对象为“test”，同时还可以看出该对象文本长度和修改时间等信息。

2. 具体任务实现

（1）为项目研发部创建公共存储容器，名为“RD_Dept_Public”

通过 Dashboard 界面创建容器。

① 进入 Dashboard 界面单击“项目”选项。

② 打开对象存储面板，单击“容器”选项。

③ 选择“创建容器”，如图 7-24 所示。

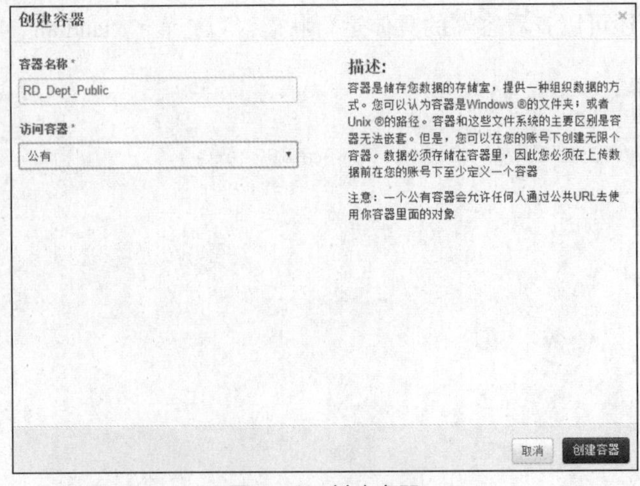

图 7-24　创建容器

④ 在弹出窗口中，输入模板源和环境源，单击“下一步”按钮。

⑤ 输入栈名、密码、镜像和网络 ID，并运行，创建后效果如图 7-25 所示。

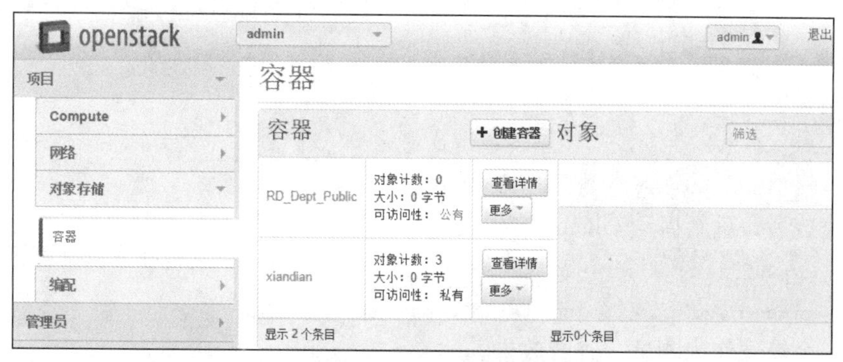

图 7-25　创建后的容器

（2）为业务部创建私有存储容器，名为“BS_Dept_Private”

通过 CLI 命令行创建容器，命令和执行结果如下。

```
# swift post BS_Dept_Private
# swift list
BS_Dept_Private
RD_Dept_Public
```

（3）为 IT 工程部创建私有存储容器，名为“IT_Dept_Private”

命令和执行结果如下。

```
# swift post IT_Dept_Private
# swift list
BS_Dept_Private
IT_Dept_Private
RD_Dept_Public
```

（4）使用 Swift 作为 Glance 的后端存储，为 Glance 存储创建名为“glance”的公共容器

```
# vi /etc/glance/glance-api.conf
[DEFAULT]
#修改后端存储方式
default_store=swift
#修改 swift keyston 版本信息
swift_store_auth_version=2
#修改 swift 认证地址
swift_store_auth_address=http://172.24.2.10:5000/v2.0/
#修改 swift 默认的租户和用户名称
swift_store_user=service:swift
#修改默认的数据密码
swift_store_key=000000
#修改 swift 存储 glance 镜像的容器名称
```

```
swift_store_container=glance
#如果存放的容器不存在，即需要创建该容器
swift_store_create_container_on_put=True
#swift 存储 glance 镜像最大容量为 5G
swift_store_large_object_size=5120
#swift 默认块大小为 200M
swift_store_large_object_chunk_size=200
```

修改完成后，重新启动 Glance 的相关服务。

```
# service openstack-glance-api restart
# service openstack-glance-registry restart
```

上传一个镜像作为测试，查看效果。

```
# glance image-create --name centos --disk-format qcow2  --container-format bare
--is-public True --progress <  /tmp/centos_65_x86_6420140327.qcow2
```

Property	Value
Checksum	dfbd01ddbb81c9e8254de236e5e83b0f
container_format	bare
created_at	2016-08-11T05:45:08
deleted	False
deleted_at	None
disk_format	qcow2
id	296b9af8-301b-4b44-b2b5-d4e558247b72
is_public	True
min_disk	0
min_ram	0
name	centos
owner	a3c430debe7e48d4a2f58c96dd89746b
protected	False
size	305397760
status	active
updated_at	2016-08-11T05:45:12
virtual_size	None

查看镜像 ID，命令和执行结果如下。

```
# glance index
ID                                      Name                    Disk Format
------------------------------------------------------------------------------
296b9af8-301b-4b44-b2b5-d4e558247b72    centos                  qcow2
81af1ee3-cf6e-4fab-8875-4077ffa95860    cirros-0.3.4-x86_64     qcow2
```

在 Swift 端执行命令查看存储信息，命令和结果如下。

```
# swift --os-username glance --os-password 000000 --os-tenant-name service
```

```
--os-auth-url
http://172.24.2.10:5000/v2.0/ list glance
296b9af8-301b-4b44-b2b5-d4e558247b72
```

（5）使用 Swift 作为 Cinder 的后端存储，为 Glance 存储创建名为“glance”的公共容器

存储节点的配置文件修改如下。

```
# vi /etc/cinder/cinder.conf
[DEFAULT]
#后端存储驱动
backup_driver=cinder.backup.drivers.swift
#Swift Endpoint 的 URL 地址
backup_swift_url=http://172.24.2.10:8080/v1/AUTH_
#Swift 认证机制
backup_swift_auth=keystone
#Swift 的用户
backup_swift_user=swift
#Swift 用户密码
backup_swift_key=000000
#默认的 Swift 存储容器
backup_swift_container=volumebackups
#备份块文件的大小
backup_swift_object_size=52428800
#备份尝试次数
backup_swift_retry_attempts=3
#重试延迟时间
backup_swift_retry_backoff=2
#压缩算法
backup_compression_algorithm=zlib
```

存储节点重启服务。

```
# service openstack-cinder-volume restart
# service openstack-cinder-backup restart
```

控制节点重启服务。

```
# service openstack-cinder-api restart
# service openstack-cinder-scheduler restart
```

创建一个作为 Cinder 后端备份存储的容器“Volume_test_backup”，执行结果如下。

```
# swift post Volume_test_backup
# swift list
BS_Dept_Private
IT_Dept_Private
RD_Dept_Public
```

```
Volume_test_backup
xiandian
```

```
# cinder list
```

ID	Status	Display Name	Size
0643885c-1dc7-4f9f-9309-382c54cc6790	available	cinder-volume-demo	1
793bc0e6-62e6-463a-b15d-24a7a13cf420	available	test	2
88952cca-11b4-48de-bb3f-d8fb7ba84bbe	available	type_test_demo	1

```
# cinder backup-create --container Volume_test_backup Volume_test
```

Property	Value
id	01ebe4fc-00f1-4ba9-a00e-35b3699aaf32
name	None
volume_id	793bc0e6-62e6-463a-b15d-24a7a13cf420

```
# cinder backup-list
```

ID	Status	Name	Size	Container
01ebe4fc-00f1-4ba9-a00e-35b3699aaf32	available	None	2	Volume_test_backup

任务总结

1. 3 种存储的对比

在学习了 Glance 镜像存储、Cinder 块存储以及 Swift 对象存储之后，3 种存储之间有什么相同点和区别？尤其作为存储来说的 Cinder 和 Swift 之间的区别又是哪些呢？

从前面几节介绍来看，总结以下几点。

Swift 对象存储是一个系统，可以上传和下载，一般存储的是不经常修改的内容，例如存储 VM 镜像、备份和归档，以及较小的文件（如照片和电子邮件消息），它更倾向于系统的管理。

Cinder 块存储具有安全可靠、高并发、大吞吐量、低时延、规格丰富、简单易用的特点，适用于文件系统、数据库或者其他需要原始块设备的系统软件或应用。

但是，这些太过于笼统，下面我们从一个简单的小例子来进行说明。

Swift 可以将 Object（理解为文件）存储到 Bucket（可以理解为文件夹）里，用 Swift 创建 Container，然后上传文件，如视频、照片，这些文件会被 Replication 到不同服务器上以保证可靠性，Swift 可以不依靠虚拟机工作。所谓的云存储，OpenStack 就是用 Swift 实现的，类似于 Amazon AWS S3（Simple Storage Service），可以理解成一个文件系统。

Cinder 是块存储（Block Storage），可以把 Cinder 当作优盘管理程序来理解。可以用 Cinder 创建 Volume，然后将它接到（Attach）虚拟机上去，这个 Volume 就像虚拟机的一个存储分区一样工作。如果把这个虚拟机 Terminate 了，这个 Volume 和里边的数据依然还在，还可以把它接到其他虚拟机上继续使用其中的数据。Cinder 创建的 Volume 必须被接到虚拟机上才能工作。类似于 Amazon AWS EBS（Elastic Block Storage），可以把 Cinder 理解成一个可移动硬盘。

而 Glance 则为虚拟机镜像提供支持，镜像资源都会存放在 Glance 存储内部。

Swift 作为一个文件系统，意味着可以为 Glance 提供存储服务，同时也可为个人的网盘应用提供存储支持。这个优势是 Cinder 和 Glance 无法实现的。

2. Swift 的应用

（1）网盘

Swift 的对称分布式架构和多 Proxy 多节点的设计导致它从基因里就适用于多用户并发的应用模式，最典型的应用莫过于类似网盘的应用。

Swift 的对称架构使得数据节点从逻辑上看处于同级别，每个节点上同时具有数据和相关的元数据，并且元数据的核心数据结构使用的是哈希环，一致性哈希算法对于节点的增减只需重定位环空间中的一小部分数据，具有较好的容错性和可扩展性。另外，数据是无状态的，每个数据在磁盘上都是完整的存储。这几点综合起来保证了存储本身的良好的扩展性。

另外，在与应用的结合上，Swift 是遵循 HTTP 的，这使得应用和存储的交互变得简单，不需要考虑底层基础构架的细节，应用软件不需要进行任何的修改就可以让系统整体扩展到非常大的程度。

（2）IaaS 公有云

Swift 在设计中的线性扩展、高并发和多租户支持等特性，使得它也非常适合作为 IaaS 的选择，公有云规模较大，更多时候会遇到大量虚拟机并发启动的情况，所以对于虚拟机镜像的后台存储来说，实际上的挑战在于大数据（超过 1GB）的并发读性能，Swift 在 OpenStack 中一开始就是作为镜像库的后台存储，经过 Rackspace 上千台机器的部署规模下的数年实践，Swift 已经被证明是一个成熟的选择。

另外，基于 IaaS 要提供上层的 SaaS 服务，多租户是一个不可避免的问题，Swift 的架构设计本身就是支持多租户的，这样对接起来更方便。

（3）备份文档

Rackspace 的主营业务就是数据的备份归档，所以 Swift 在这个领域也是久经考验的，同时，他们还延展出一种新业务——“热归档”。由于长尾效应，数据可能被调用的时间窗越来越长，“热归档”能够保证应用归档数据在分钟级别重新获取，和传统磁带机归档方案中的数小时相比，是一个很大的进步。

（4）移动互联网和 CDN

移动互联网和手机游戏等产生的大量用户数据，数据量虽然不是很大，但是用户数很多，这也是 Swift 能够处理的领域。

至于加上 CDN，如果使用 Swift，云存储就可以直接响应移动设备，不需要专门的服务器去响应这个 HTTP 的请求，也不需要在数据传输中再经过移动设备上的文件系统，而是直接使用 HTTP 上传云端。如果把经常被平台访问的数据缓存起来，利用一定的优化机制，数据可以从不同的地点分发到用户那里，这样就能提高访问的速度。在 Swift 的开发社区上有人在讨论视频网站应用和 Swift 的结合，个人以为是值得关注的方向。

项目八 高级控制服务

在学习了基本的控制服务之后，我们了解了云平台所需的基本组件。本项目继续学习辅助平台管理和满足生产要求的一些高级的平台组件，以提高平台快速部署实例的应用能力及平台的监控管理能力。

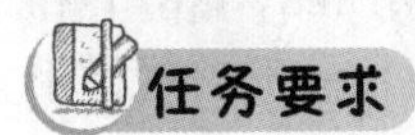

学习目标

- 了解 Heat 组件的功能、架构及使用。
- 掌握 Heat 模板的编写方法。
- 掌握编配服务、栈和模板的关键概念。
- 掌握使用辅助 Shell 脚本完成对平台的监控。

任务一 编配服务

任务要求

Heat 模板文件可以实现实例的批量自动部署及应用，保证用户可以正常使用云平台的实例资源。编写 Heat 模板文件时可根据需要选择相应的镜像和网络。

小李要为项目研发部、业务部和 IT 工程部编写 Heat 模板文件，该模板文件可以启动 3 台实例，其中一台为 Win 7 64 位系统、4GB 内存和 60GB 磁盘空间，另外两台为 CentOS 6.5 64 位操作系统、1GB 内存和 50GB 磁盘空间。

相关知识

1. 基本概念

Heat 是一个基于模板来编排复合云应用的服务。它目前支持亚马逊的 CloudFormation 模板格式，也支持 Heat 自有的 Hot 模板格式。模板的使用简化了复杂的基础设施、服务和应用的定义和部署。模板支持丰富的资源类型，不仅覆盖了常用的基础架构，包括计算、网络、存储、镜像，还覆盖了 Ceilometer 的警报、Sahara 的集群和 Trove 的实例等高级资源。它可以基于模板来实现云环境中资源的初始化、依赖关系处理、部署等基本操作，也可以解决自动收缩、负载均衡等高级特性。目前，Heat 自有的 Hot 模板格式正在不断地改进，同时也支持 AWS CloudFormation 模板（CFN）。下面我们将详细讲解 Hot 模板。

Heat 服务包含以下重要的组件。

① Heat API 组件可实现 OpenStack 天然支持的 REST API。该组件通过把 API 请求经

由 AMQP 传送给 Heat Engine 来处理 API 请求。

② Heat API CFN 组件提供兼容 AWS CloudFormation 的 API，同时也会把 API 请求通过 AMQP 转发给 Heat Engine。

③ Heat Engine 组件提供 Heat 最主要的协作功能。

首先，用户在 Horizon 中或者命令行中提交包含模板和参数的输入请求，Horizon 或者命令行工具会把请求转化为 REST 格式的 API 并调用 Heat API 或 Heat API CFN。然后 Heat API 和 Heat API CFN 会验证模板的正确性，并通过 AMQP 异步传递给 Heat Engine 来处理请求，如图 8-1 所示。

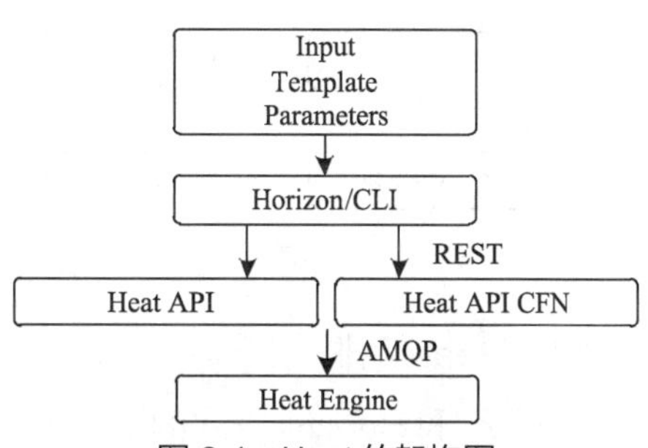

图 8-1　Heat 的架构图

当 Heat Engine 拿到请求后，会把请求解析为各种类型的资源，每种资源都对应 OpenStack 的其他服务客户端，然后发送 REST 请求给其他服务。通过如此的解析和协作，最终完成对请求的处理。

在这里，Heat Engine 的作用分为 3 层：第 1 层处理 Heat 层面的请求，就是根据模板和输入参数来创建 Stack，这里的 Stack 由各种资源组合而成。第 2 层解析 Stack 里各种资源的依赖关系，以及 Stack 和嵌套 Stack 的关系。第 3 层就是根据解析出来的关系，依次调用各种服务客户端来创建各种资源，如图 8-2 所示。

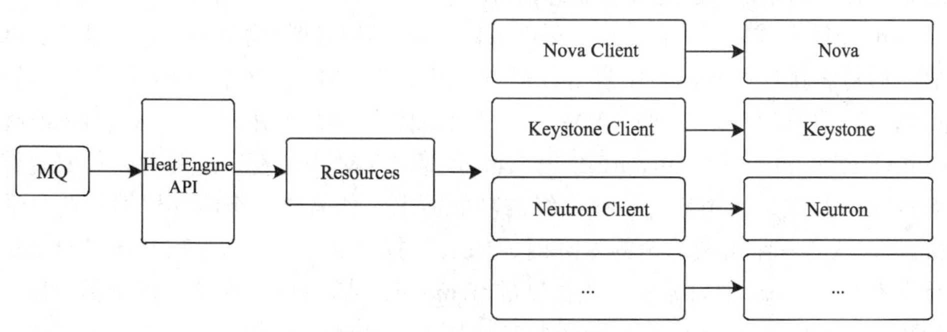

图 8-2　Heat Engine 结构

2. 编排

编排，顾名思义，就是按照一定的目的依次排列。在 IT 世界，一个完整的编排一般包括，设置服务器上的机器，安装 CPU、内存、硬盘，通电，插入网络接口，安装操作系统，配置操作系统，安装中间件，配置中间件，安装应用程序以及配置应用发布程序。对于复杂的需要部署在多台服务器上的应用，需要重复这个过程，而且需要协调各个应用模块的配置。例如，配置前面的应用服务器连接上后面的数据库服务器。图 8-3 显示了一个典型应用需要编排的项目。

在云计算的世界里，机器这层就变为虚拟的 VM 或者是容器。管理 VM 所需要的各个资源要素和操作系统本身就成了 IaaS 编排的重点。操作系统安装完后的配置也是 IaaS 编排所覆盖的范围。除此之外，提供能够接入 PaaS 和 SaaS 编排的框架也是 IaaS 编排的范围。

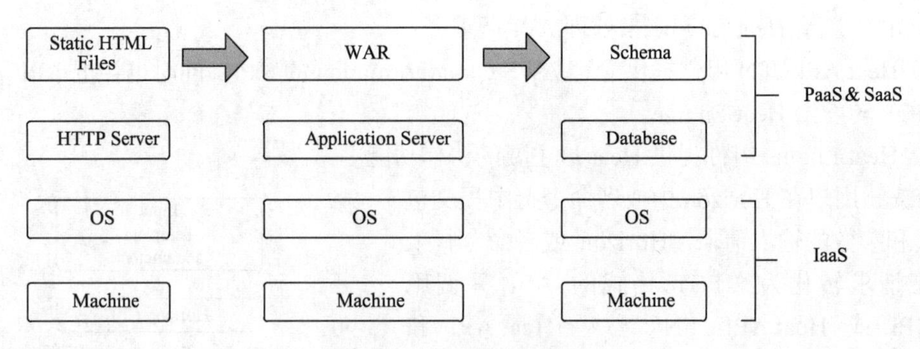

图 8-3 应用项目的编排图

3. Heat 编排

命令行和 Horizon 方式的工作效率并不高，即使把命令行保存为脚本，在 I/O、依赖关系之间仍需要编写额外的脚本来进行维护，而且不易于扩展。如果用户直接通过 REST API 编写程序，同样会引发额外的复杂性。因此，这两种方式都不利于用户通过 OpenStack 进行批量资源管理和编排各种资源。

Heat 在这种情况下应运而生，它采用了业界流行的模板方式进行设计和定义编排。用户只需要打开文本编辑器，编写一段基于 Key-value 的模板，就能够方便地得到想要的编排。为了方便用户的使用，Heat 提供了大量的模板例子，通常用户只需要选择想要的编排，通过复制、粘贴的方式就能完成模板的编写。

Heat 的编排方式如下。首先，OpenStack 自身提供基础架构资源，包括计算、网络和存储等。通过编排这些资源，用户可以得到最基本的 VM。此外，在编排 VM 的过程中，用户可以编写简单脚本，以便对 VM 做简单的配置。然后，用户可以通过 Heat 提供的 Software Configuration 和 Software Deployment 等对 VM 进行复杂的配置，如安装软件和配置软件等。其次，当用户有一些高级的功能需求，如需要一组能够根据负荷自动伸缩的 VM 组，或者一组负载均衡的 VM 时，Heat 可提供 Auto Scaling 和 Load Balance 等模板进行支持。在 Heat 中只需要一段长度的 Template，就可以实现这些复杂的应用。Heat 对诸如 Auto Scaling 和 Load Blance 等复杂应用的支持已经非常成熟，有各种各样的模板可供参考。

4. Heat 模板

目前，Heat 支持两种格式的模板，一种是基于 JSON 格式的 CFN 模板，另一种是基于 YAML 格式的 HOT 模板。CFN 模板主要是为了保持对 AWS 的兼容性。HOT 模板是 Heat 自有的，资源类型更加丰富，更能体现出 Heat 的特点。

一个典型的 HOT 模板由下列元素构成。

（1）模板版本

必填，指定所对应的模板版本，Heat 会根据版本进行检验。

（2）参数列表

选填，指输入参数列表。

（3）资源列表

必填，指生成的 Stack 所包含的各种资源。可以定义资源间的依赖关系，例如生成 Port，然后再用 Port 来生成 VM。

（4）输出列表

选填，指生成的 Stack 暴露出来的信息，可以用来给用户使用，也可以用来作为输入提供给其他的 Stack。

举例如下。

```
heat_template_version: 2013-05-23        //版本信息
parameters:                              //自定义的变量
ImageID:
type: string
resources:                               //描述的资源，如网络、实例
 server1:
type: OS::Nova::Server
outputs:                                 //返回值
 server1_private_ip:
value: { get_attr: [ server1, first_address ] }
```

对于不同的资源，Heat 都提供了对应的资源类型，可以通过 Heat 的命令进行查询。

```
# heat resource-type-list
+------------------------------------------+
| resource_type                            |
+------------------------------------------+
| AWS::CloudFormation::Stack               |
| AWS::EC2::EIP                            |
| AWS::ElasticLoadBalancing::LoadBalancer  |
| AWS::IAM::AccessKey                      |
| AWS::IAM::User                           |
| AWS::RDS::DBInstance                     |
| AWS::S3::Bucket                          |
| OS::Ceilometer::Alarm                    |
| OS::Ceilometer::CombinationAlarm         |
| OS::Cinder::Volume                       |
| OS::Cinder::VolumeAttachment             |
| OS::Neutron::LoadBalancer                |
| OS::Neutron::Net                         |
| OS::Neutron::NetworkGateway              |
| OS::Neutron::Pool                        |
| OS::Neutron::Port                        |
| OS::Neutron::Subnet                      |
| OS::Nova::FloatingIP                     |
| OS::Swift::Container                     |
| OS::Trove::Instance                      |
+------------------------------------------+
```

例如，对于 VM，Heat 提供了“OS::Nova::Server”命令。“OS::Nova::Server”命令有一些参数，如 key、image 和 flavor 等，这些参数可以直接指定，可以由客户在创建 Stack 时提供，也可以由上下文其他的参数获得。

举例如下。

```
//Heat 资源使用及其参数设定
resources:
private_net:
type: OS::Neutron::Net
properties:
name: private-net
private_subnet:
type: OS::Neutron::Subnet
properties:
network_id: { get_resource: private_net }
cidr: 172.16.2.0/24
gateway_ip: 172.16.2.1
```

对于 properties 中的参数形式，以网络为例，可以从已有的网络中获取。

```
# neutron subnet-show IT-Subnet       //查看 IT-subnet 子网详细信息
```

Field	Value
allocation_pools	{"start": "172.24.5.2", "end": "172.24.5.254"}
cidr	172.24.5.0/24
dns_nameservers	
enable_dhcp	True
gateway_ip	172.24.5.1
host_routes	
ids	82b79434-66f6-4743-aa98-1a69ba2877c6
ip_version	4
name	IT-Subnet
network_id	3f01d141-b802-41bd-b429-8832c4c5b00d
tenant_id	219c95eac5694e45bd5c7304613835d3

Heat 提供了多种资源类型来支持对软件的配置和部署的编配，也就是说，通过 Heat 启动实例的时候可以在实例中预装软件，如下。

```
OS::Heat::CloudConfig:
//VM 引导程序启动时的配置，由"OS::Nova::Server"引用
OS::Heat::SoftwareConfig:
//描述软件配置
OS::Heat::SoftwareDeployment:
//执行软件部署
OS::Heat::SoftwareDeploymentGroup:
```

```
//对一组 VM 执行软件部署
OS::Heat::SoftwareComponent:
//针对软件的不同生命周期部分，对应描述软件配置
OS::Heat::StructuredConfig:
//和 OS::Heat::SoftwareConfig 类似，但是用 Map 来表述配置
OS::Heat::StructuredDeployment:
//执行 OS::Heat::StructuredConfig 对应的配置
OS::Heat::StructuredDeploymentsGroup:
//对一组 VM 执行 OS::Heat::StructuredConfig 对应的配置
//举例：通过 OS::Heat::SoftwareConfig 在实例中安装 FTP 服务
resources:
install_ftp_serviceconfig
type: OS::Heat::SoftwareConfig
properties:
group: script
outputs:
- name: result
config: |
#!/bin/bash -v
yum -y vsftpd
sed -i '$a anon_root=/opt' /etc/vsftpd/vsftpd.conf
servicevsftpd start
chkconfigvsftpd on
```

Heat 和 Ceilometer 服务可以实现对资源的自动伸缩编配，还可以实现对资源的负载均衡编配。未来，Heat 在 OpenStack 中将完成更多重要功能，如所谓的弹性扩展等。现下最热门的 App Store，也是由 Heat 来实现的。

任务实现

1. Heat 的运维基础

（1）使用栈模板“test-stack.yml”创建一个名为“Orchestration”的栈

命令和执行结果如下。

```
# heat stack-create orchestration -f test-stack1.yml --parameters "ImageID=
centos6.5;NetID=sharednet1"
```

id	Stack_name	Stack_status	Creation_time
25d3b7cc-a107-4003-aafa-52b561972372	RD_Dept	CREATE_COMPLETE	2016-04-13T02:44:15Z
1023afdb-741d-4059-ae6b-9e196584dffd	BD_Dept	CREATE_COMPLETE	2016-04-13T03:03:47Z
f850fa49-dacc-4d85-8e6d-1c28f0975212	IT_Dept	CREATE_COMPLETE	2016-04-13T03:16:01Z
1ca61660-3809-4f90-9f1c-36ac60157aa6	Guest	CREATE_COMPLETE	2016-04-13T03:19:26Z
6cf70911-034b-4a1e-b0b6-18198d16950f	orchestration	CREATE_IN_PROGRESS	2016-04-15T07:17:55Z

（2）查看栈列表

命令和执行结果如下。

```
# heat stack-list
```

id	Stack_name	Stack_status	Creation_time
25d3b7cc-a107-4003-aafa-52b561972372	RD_Dept	CREATE_COMPLETE	2016-04-13T02:44:15Z
1023afdb-741d-4059-ae6b-9e196584dffd	BD_Dept	CREATE_COMPLETE	2016-04-13T03:03:47Z
f850fa49-dacc-4d85-8e6d-1c28f0975212	IT_Dept	CREATE_COMPLETE	2016-04-13T03:16:01Z
1ca61660-3809-4f90-9f1c-36ac60157aa6	Guest	CREATE_COMPLETE	2016-04-13T03:19:26Z
6cf70911-034b-4a1e-b0b6-18198d16950f	orchestration	CREATE_IN_PROGRESS	2016-04-15T07:17:55Z

（3）查看栈的详细信息

```
# heat stack-show orchestration
```

（4）删除栈

```
# heat stack-delete orchestration
```

（5）查看栈资源列表

命令和执行结果如下。

```
# heat resource-list orchestration
```

resource_name	physical_resource_id	resource_type	resource_status	updated_time
Server1	4934dd31-441b-42f7-Becf-8f2be96cc829	OS::Nova::Server	CREATE_COMPLETE	2016-04-15T07:17:55Z
server2	57cb0972-ddf2-4096-bdc4-0598bb943d7f	OS::Nova::Server	CREATE_COMPLETE	2016-04-15T07:17:55Z

（6）查看栈资源

命令和执行结果如下。

```
# heat resource-show orchestration server1
```

Property	Value
description	
links	http://58.214.31.6:8004/v1/b9b2b3bf6fef409497a21e0c4860aed7/stacks/openstac http://58.214.31.6:8004/v1/b9b2b3bf6fef409497a21e0c4860aed7/stacks/openstac
logical_resource_id	server1
physical_resource_id	2f1526e5-5aec-429x-9285-5589d6c2b1c1
required_by	
resource_name	Server1
resource_status	CREATE_COMPLETE
resource_status_reason	state changed
resource_type	OS::Nova::Server
updated_time	2016-07-21T05:05:34Z

（7）查看输出列表

命令和执行结果如下。

```
# heat output-list orchestration
```

output_key	description
server1_private_ip	Ip address of the server in the private network
server2_private_ip	Ip address of the server in the private network

（8）查看输出值

```
# heat output-show orchestration server1_private_ip
"172.24.4.2"
```

（9）查看事件列表

命令和执行结果如下。

```
# heat event-list orchestration
```

resource_name	id	resource_status_reason	resource_status	event_time
Server1	10c307a5-1732-4f10-8b91-d9b23402661d	state changed	CREATE IN PROGRESS	2016-04-15T07:17:55Z
server2	2036acec-8fSa-4951-8060-aa6999ece9fe	state changed	CREATE IN PROGRESS	2016-04-15T07:17:57Z
Server1	744f8fff-17b4-452e-b37a-b7b147090d05	state changed	CREATE_COMPLETE	2016-04-15T07:18:10Z
Server2	c664f1d5-7e9a-48da-aa94-707a187f941d	state changed	CREATE_COMPLETE	2016-04-15T07:18:13Z

（10）查看资源事件详细信息

命令和执行结果如下。

```
# heat event-show orchestration server1 10c307a5-1732-4f10-8b91- d9b23402661d
```

Property	Value
event_time	2016-07-21T05:05:34Z
id	10c307a5-1732-4f10-8b91-d9b23402661d
links	http://58 214 31.6 8004/vl/b9b2b3bf6fef409497a21c0c4860aed7/stacks/openstac http://58.214.31.6:8004/vl/b9b2b3bf6fef409497a21c0c4860aed7/stacks/openstac http://58.214.31.6:8004/vl/b9b2b3bf6fef409497a21e0c4860aed7/stacks/openstac
logical_resource_id	server1
physical_resource_id	None
resource_name	server1
resource_properties	{

```
"admin_pass"=nulll,
"user_data_format": "HEAT_CFNTOOLS",
"admin_user": null,
"name": "teacher61",
"block_device_mapping": null,
"key_name": null,
"image": "OpenStack_already",
"availability_zone": null,
"image_update_policy": "REPLACE",
"software_config_transport": "POLL_SERVER_CFN",
"diskConfig": null,
"metadata": null,
"personality": {},
"user_data": "",
"flavor_update_policy": "RESIZE",
"flavor": "OpenStack",
"config_drive": null,
"reservation_id": null,
"networks": [
  {
    "uuid": null,
    "fixed_ip": null,
    "network": "network91",
    "port": null
  }
 ],
 "security_groups": [],
 "scheduler_hints": null
}
CREATE_IN_PROGRESS
state changed
OS::Nova::Server
```

2. 完成编配服务任务

（1）编写 RD_Dept.yml 文件

```
//编写 Dept_Server_3.yml 文件，栈文件中的参数指定虚拟机的数量、镜像和网络等。
# vi Dept_Server_3.yml
heat_template_version: 2013-05-23
description: Test Template
parameters:
  Net:
```

```
type: string
description: Network for the server
resources:
  server1:
type: OS::Nova::Server
properties:
name: "Cloud_Win7_x64 server1"
image: "Cloud_Win7_x64"
flavor: "m1.medium"
networks:
     - network: { get_param: Net }
  server2:
type: OS::Nova::Server
properties:
name: "Cloud_Centos6.5_64 server1"
image: Cloud_Centos6.5_64
flavor: "m1.small"
networks:
     - network: { get_param: Net }
  server3:
type: OS::Nova::Server
properties:
name: "Cloud_Centos6.5_64 server2"
image: Cloud_Centos6.5_64
flavor: "m1.small"
networks:
     - network: { get_param: Net }
outputs:
  server1_private_ip:
description: IP address of the server in the private network
value: { get_attr: [ server1, first_address ] }
  server2_private_ip:
description: IP address of the server in the private network
value: { get_attr: [ server2, first_address ] }
  server3_private_ip:
description: IP address of the server in the private network
value: { get_attr: [ server3, first_address ] }
```

（2）通过 Dashboard 界面为项目研发部启动栈资源，创建实例

① 进入 Dashboard 界面找到“项目”选项。

② 打开编配面板，找到“栈”。

③ 选择“启动栈”。

④ 在弹出窗口中，输入“模板源”和“环境源”，单击“下一步”按钮，如图 8-4 所示。

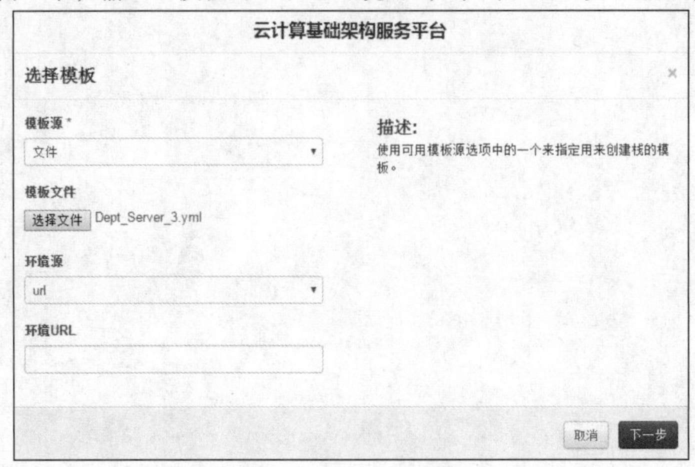

图 8-4　为实例选择模板

⑤ 输入“栈名”“用户‘admin’的密码”和“Net”，单击“运行”按钮，如图 8-5 所示。

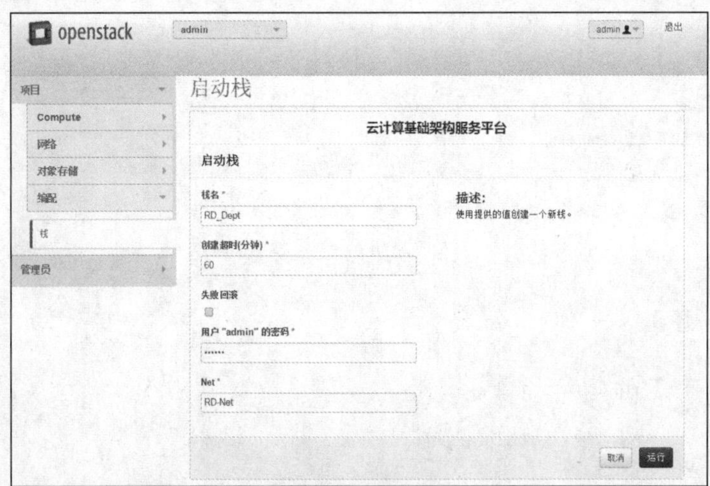

图 8-5　启动栈

⑥ 创建成功后，查看栈的信息，如图 8-6 所示。

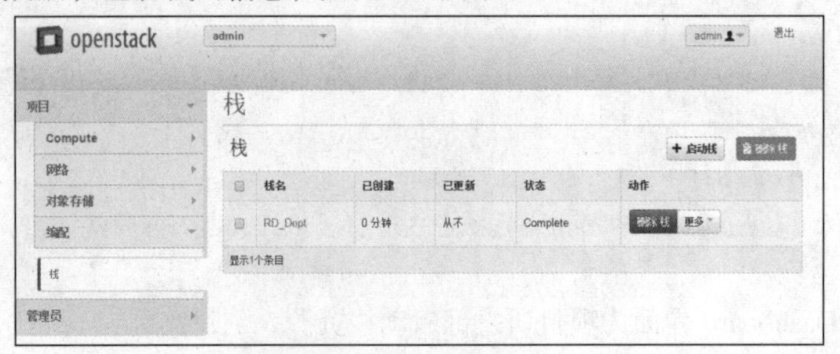

图 8-6　查看栈的信息

⑦ 查看创建栈的实例，如图 8-7 所示。

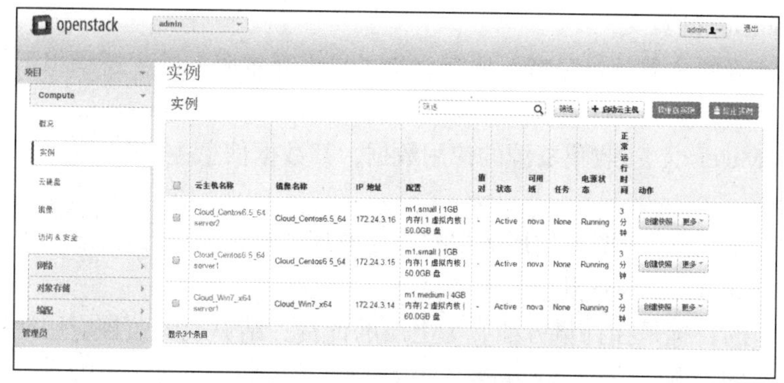

图 8-7　查看创建栈的实例

（3）通过 Shell 命令行为业务部启动栈资源，创建实例

命令和执行结果如下。

```
# heat stack-create -f Dept_Server_3.yml --parameters "Net=BS-Net" BS_Dept
```

id	Stack_name	Stack_status	creation_time
82c84043-57dd-48ba-8049-9587659ed6d8	BS_Dept	CREATE_IN_PROGRESS	2016-08-12T06:33:49Z

（4）通过 Shell 命令行为 IT 工程部启动栈资源，创建实例

命令和执行结果如下。

```
# heat stack-create -f Dept_Server_3.yml --parameters "Net=IT-Net" IT_Dept
```

id	Stack_name	Stack_status	creation_time
82c84043-57dd-48ba-8049-9587659ed6d8	BS_Dept	CREATE_FAILED	2016-08-12T06:33:49Z
b4f41c26-a8ae-4ce6-8107-33dec6a01f39	IT_Dept	CREATE_IN_PROGRESS	2016-08-12T06:34:52Z

任务二　监控服务

任务要求

接下来，小李要熟悉监控服务。通过监控服务，可以实时查看平台的运行情况，保障平台运行稳定，维护数据安全，对可能出现的危险做到快速判断和处理。

要求可以通过 GUI 或 CLI 查看平台某一个时间段的运行数据，包括网络数据、实例数据、存储数据和服务资源消耗情况数据。

相关知识

一般来说，云计算平台除了提供服务外，还担任了“计量、计费”等功能。本节将详细介绍云测量服务 Ceilometer。

1. 基本概念

（1）Ceilometer 的意义

通常来说，云服务尤其是公有云服务，除了提供基本服务外，还承担了“计费”的功能，公有云在计费方面有 3 个层次。

① 计量（Metering）：收集资源的使用数据，其数据信息主要包括使用对象、使用者、使用时间和用量。

② 计费（Rating）：将资源使用数据按照商务规则转化为可计费项目并计算费用。

③ 结算（Billing）：收钱开票。

Ceilometer 的目标是在计量方面，为上层的计费、结算或者监控应用使用数据收集功能，以提供统一的资源。

（2）Ceilometer 主要概念

Ceilometer 的主要概念包括以下 5 类。

① Meter：是资源使用的计量项，它的属性包括名称（Name）、单位（Unit）、类型（Cumulative 表示累计值、Delta 表示变化值、Gauge 表示离散或者波动值）以及对应的资源属性等。

② Sample：是某个时刻某个资源（Resource）的某个 Meter 值。Sample 的集有区间概念，即收集数据的时间间隔。除了 Meter 属性外，还有 Timestamp（采样时间）和 Volume（采样值）属性。

③ Statistics：一个时间段（Period）内的 Samples 聚合值，包括计数（Count）、最大（Max）、最小（Min）、平均（Avg）、求和（Sum）等。

④ Resource：指被监控的资源对象，可以是一台虚拟机、一台物理机或者一块云硬盘。

⑤ Alarm：是 Ceilometer 的告警机制，可以通过阈值或者组合条件告警，并设置告警时触发的 Action。

（3）Alarm 的状态

```
ALARM （告警状态）：
{"current": "alarm", "alarm_id": "742873f0-97f0-4d99-87da-b5f7c7829b7f", "reason": "Remaining as alarm due to 1 samples outside threshold, most recent: 0.138333333333", "previous": "alarm"}
正常状态 （数据充足，未告警）：
{"current": "ok", "alarm_id": "742873f0-97f0-4d99-87da-b5f7c7829b7f", "reason": "Remaining as ok due to 1 samples inside threshold, most recent: 0.138333333333", "previous": "ok"}
Insufficient Data （默认状态，数据不足）：
{"current": "insufficient data", "alarm_id": "742873f0-97f0-4d99-87da-b5f7c7829b7f", "reason": "1 datapoints are unknown", "previous": "ok"}
```

（4）Alarm 的动作

目前 Ceilometer 支持两种 Action。

① 'log:'：Alarm 被写入 Log 文件中。

② 'WebhookURL:'：这是一个 HTTP(S) Endpoint 的 URL，如 'http://130.56.250.199:8080/alarm/instances_TOO_MANY'。Alarm 的内容会以 JSON 的格式被 POST 到该 URL 中。

（5）Ceilometer 采集机制

在 Ceilometer 的各个服务中，与采集相关的服务是 Ceilometer Collector、Ceilometer Agent-Central、Ceilometer Agent-Compute 和 Ceilometer Agent-Notification。具体流程如图 8-8 所示。

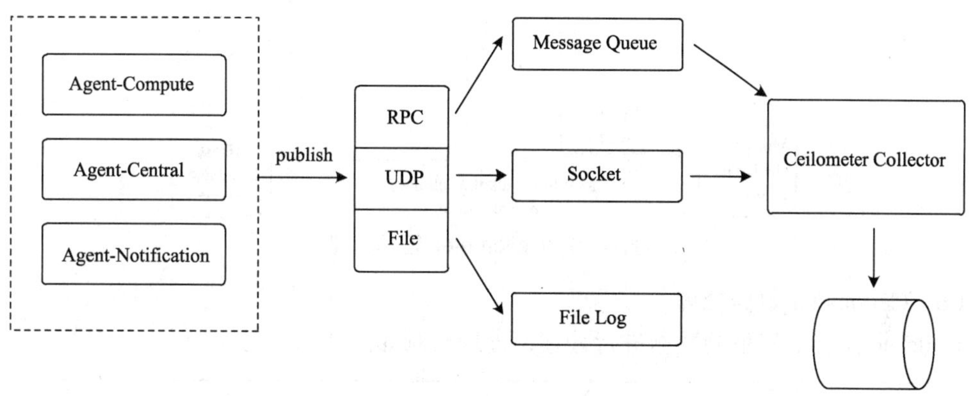

图 8-8　Ceilometer 采集机制

Agent-*服务负责采集信息，采集的信息可以通过 3 种方式 Publish 出来，包括 RPC、UDP 和 File。RPC 是将采集的信息以 Payload 方式发布到消息队列；Collector 服务通过监听对应的 Queue 来收集这些信息，并保存到存储介质中；UDP 通过 Socket 创建一个 UDP 数据通道，然后 Collector 通过 Bind 这个 Socket 来接收数据，并保存到存储介质中；File 方式比较直接，就是将采集的数据以 Filelog 的方式写入 Log 文件中。

至于使用哪种方式 Publish，要看 Pipeline 文件是如何配置的，具体可以查看 /etc/ceilometer/pipeline.yaml 中的 Publishers 配置。

Agent-*三个采集组件分别负责采集不同类型的信息，Agent-Notification 负责收集各个组件推送到 Oslo Messaging 的消息；Oslo Messaging 是 OpenStack 整体的消息队列框架，所有组件的消息队列都使用这个组件；Agent-Compute 只负责收集虚拟机的 CPU 内存和 I/O 等信息，所以需要安装在 Hypervisor 机器上；Agent-Central 通过各个组件的 API 方式收集有用的信息；Agent-Notification 只需监听 AMQP 中的 Queue 即可收到信息；Agent-Compute 和 Agent-Central 都需要定期轮询收集信息，如图 8-9 所示。

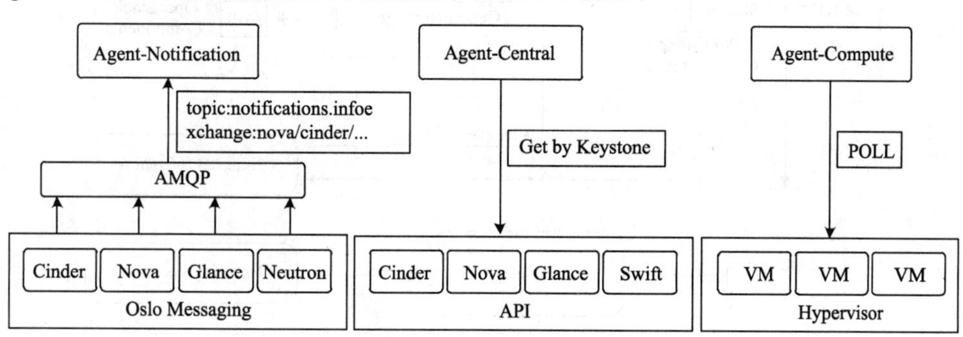

图 8-9　定期轮询收集信息

信息通过 Agent-*采集并由 Collector 汇总处理，最终需要持久化到存储介质中，Ceilometer 目前支持的存储包括 MySQL、DB2、HBase 和 MongoDB，从支持的数据库来看，监控数据持久化的压力还是相当大的，如图 8-10 所示。

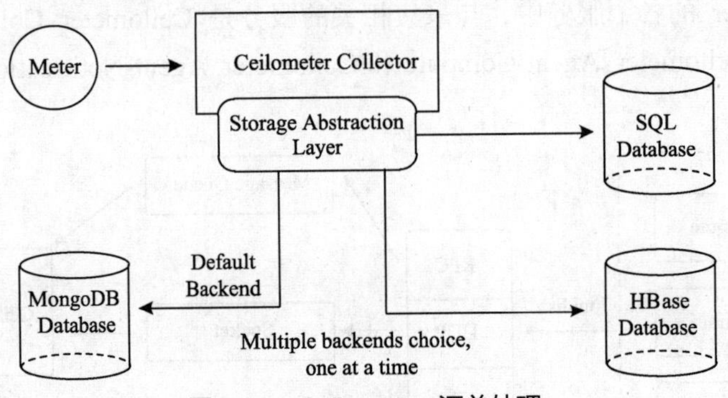

图 8-10　Ceilometer 汇总处理

（6）Ceilometer 数据处理

Ceilometer 会对采集的数据进行处理，处理机制如图 8-11 所示。

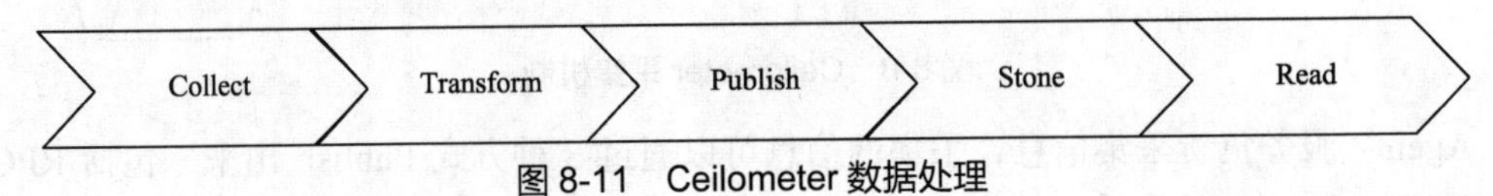

图 8-11　Ceilometer 数据处理

① Collect：Meters 数据收集。

② Transform：Meters 数据转换。

③ Publish：Meters 数据发布。

④ Store：Meters 数据保存。

⑤ Read：Meters 数据访问。

⑥ Alarm：提供告警。

总体架构如图 8-12 所示。

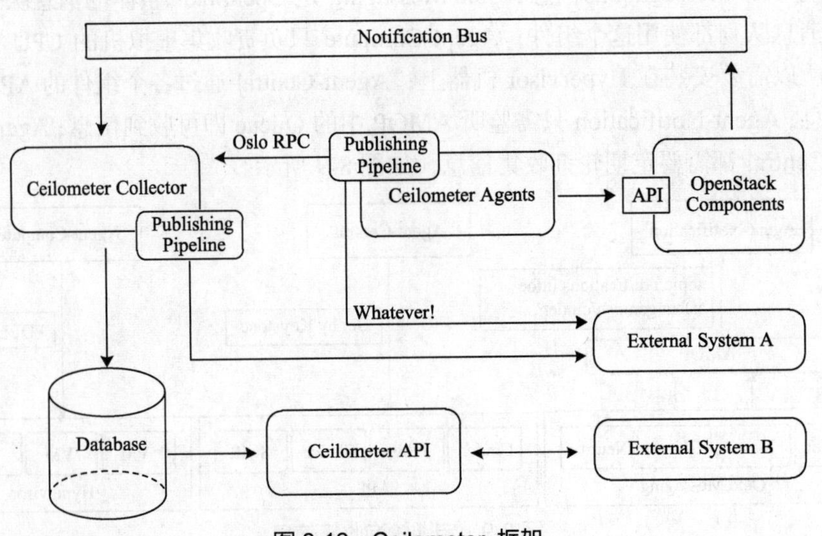

图 8-12　Ceilometer 框架

2. Meter 的数据处理

Meters 的数据处理使用的是 Pipeline 的方式，即 Metes 数据依次经过（零个或者多个）Transformer 和（一个或者多个）Publisher 处理，最后达到（一个或者多个）Receiver。其中，Recivers 包括 Ceilometer Collector 和外部系统。

Ceilometer 根据配置文件/etc/ceilometer/pipeline.yaml 来配置 Meters 所使用的 Transformers 和 Publishers，如图 8-13 所示。下面以 cpu meter 为例介绍 Pipeline 配置。

```
sources: A source is a producer of samples//对 cpu 的 meters 数据进行处理，并进行数据取样
......
  - name: cpu_source
interval: 600
meters:
       - "cpu"
sinks:
       - cpu_sink
......
sinks: A sink on the other hand is a chain of handlers of samples
......
  - name: cpu_sink
transformers:
     - name: "rate_of_change"
parameters:
target:
name: "cpu_util"
unit: "%"
type: "gauge"
scale: "100.0 / (10**9 * (resource_metadata.cpu_number or1))"
publishers:
       - notifier://
```

这段代码定义了 cpu meter 的一些属性。

① “interval: 600” 表示 Poller 获取 cpu samples 的间隔为 10 分钟。

② cpu meter 的 Transformer 为 “rate_of_change”。

③ cpu meter 的 Publisher 为 notifier://，它使用默认的配置，并经过 AMQP 使用 oslo.messaging 发出数据。

Transformer，即 Sample 的转换器，常见的 Transformer 如下。

① unit_conversion：单位转换器，如温度从°F 转换成℃。

② rate_of_change：计算方式转换器，如根据一定的计算规则来转换一个 Sample。

③ accumulator：累计器，对测量转换过的结果进行累加或迭代操作。

例如：

```
name: "rate_of_change"        //数据处理动作为数据改变
   parameters:
   target:
name: "cpu_util"
unit: "%"
type: "gauge"
scale: "100.0 / (10**9 * (resource_metadata.cpu_number or 1))"
```

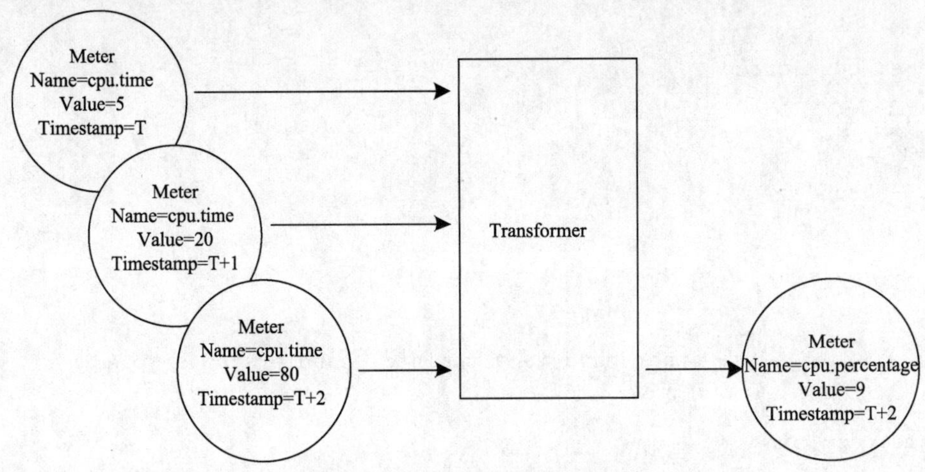

图 8-13　Ceilometer 数据处理

3. Publisher 分发器

Ceilometer 支持的 Publisher 类型如表 8-1 所示。

表 8-1　Publisher 类型

Publishers 类型	格　式	说　明	配置项
Notifier	notifier://?option1=value1&option2=value2	Samples 数据被发送到 AMQP 系统，然后被 Ceilometer Collecter 接收。默认的 AMQP Queue 是 metering_topic=metering。这是默认的方式	[publisher_notifier] metering_driver=messagingv2 metering_topic = metering
RPC	rpc://?option1=value1&option2=value2	与 Notifier 类似，同样经过 AMQP，不过是同步操作，因此可能有性能问题	[publisher_rpc] metering_topic = metering
UDP	udp://<host>:<port>/	经过 UDP Port 发出。默认的 UDP 端口是 4952	udp_port=4952
File	file://path?option1=value1&option2=value2	发送到文件保存	

通过修改/ect/ceilometer/ceilometer.conf 可配置 Publisher 分发器，以下列出了集中 Publisher 的配置项以及消息示例。

Notifier 配置项如下。

```
[publisher_notifier]
metering_driver = messagingv2
metering_topic = metering
```

示例：

```
notifier://?policy=drop&max_queue_length=512
```

RPC 配置项如下。

```
[publisher_rpc]
metering_topic = metering
```

示例：

```
rpc://?per_meter_topic=1
```

UDP 配置项如下。

```
udp_port=4952
```

示例：

```
rpc://?per_meter_topic=1
```

File 配置项如下。

```
file://path?option1=value1&option2=value2
```

通过配置/etc/ceilometer/pipeline.yaml 来为某个 Meter 指定多个 Publisher。例如，增加一个 UDP 的 Publisher。

```
sinks:
   - name: meter_sink
transformers:
publishers:
        - notifier://
      - rpc://?per_meter_topic=1
```

4. 数据保存

（1）Ceilometer Collector 从 AMQP 接收到数据后，会原封不动地通过一个或者多个分发器（Dispatchers）将它保存到指定位置，如图 8-14 所示。目前它支持的分发器如下。

① 文件分发器：保存到文件。

通过添加配置项“dispatcher = file ”来指定分发器类型为文件。

② HTTP 分发器：保存到外部的 HTTP Target。

通过添加配置项“dispatcher = http”来指定分发器类型为 HTTP。

③ 数据库分发器：保存到数据库。

添加配置项“dispatcher = database”来指定分发器类型为数据库。

（2）Ceilometer Collector 支持的数据库类型分发器有以下 3 种。

① MongoDB：默认 DB。

② SQL DB：支持 MySQL、PostgreSQL 和 IBM DB2 等。

③ HBase DB。

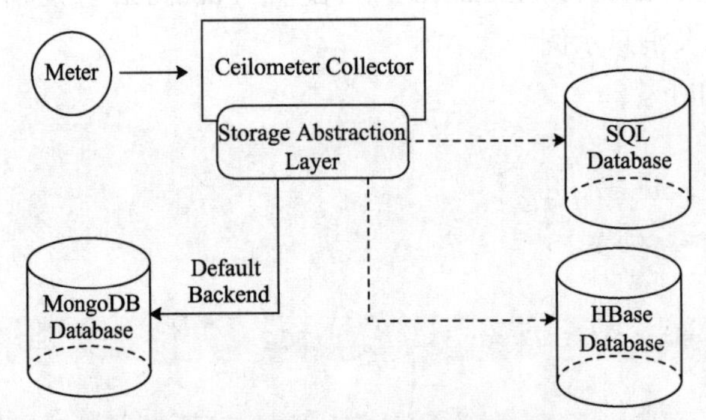

图 8-14 Ceilometer 数据库类型

Ceilometer 支持同时配置多个分发器，将数据保存到多个目的位置。通过在 ceilomet-er.conf 中做如下配置，可以同时使用 File 和 Database Dispatcher。

```
[DEFAULT]
dispatcher = database
dispatcher = file
[dispatcher_file]
backup_count = 5
file_path = /var/log/ceilometer/ceilometer-samples
max_bytes = 100000
```

5. 告警

Ceilometer Alarm API 使用 Ceilometer REST API 获取 Statistics 数据。

Ceilometer Alarm Evaluator 可生成 Alarm 数据，并通过 AMQP 发给 Ceilometer Alarm Notifer。

Ceilometer Alarm Notifer 会通过指定方式把 Alarm 发出去。具体流程如图 8-15 所示。

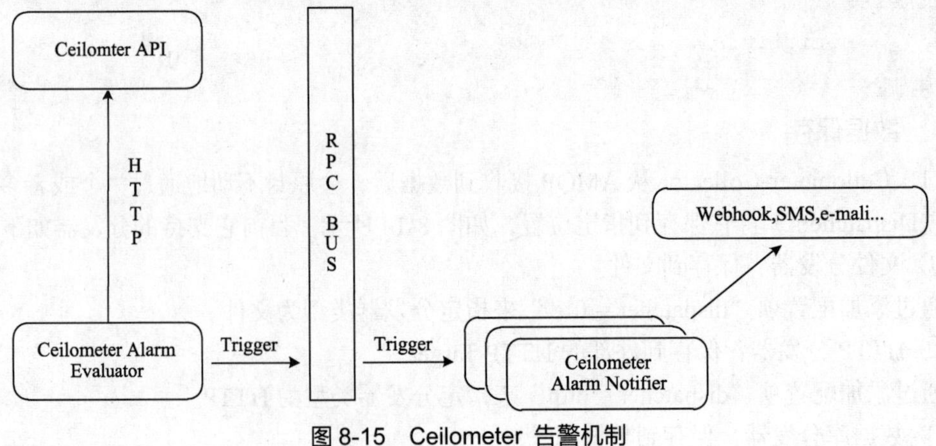

图 8-15 Ceilometer 告警机制

任务实现

1. 数据查看

要求可以通过 GUI 或 CLI 查看平台某一个时间段的运行数据，包括网络数据、实例数据、存储数据和服务资源消耗情况数据。

① 进入 Dashboard 界面单击“管理员”选项。

② 打开系统面板，单击“资源使用情况”。

③ 进入到“统计数据”界面，如图 8-16 所示。

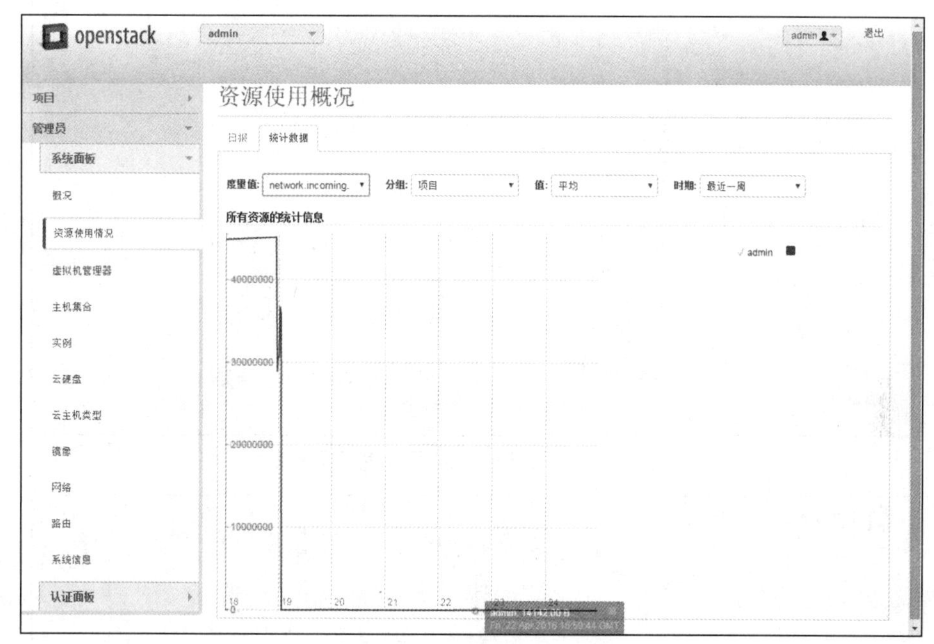

图 8-16　查看资源使用情况

④ 选择相应的“度量值”“分组”“值”“时期”，查看统计信息，如图 8-17、图 8-18 和图 8-19 所示。

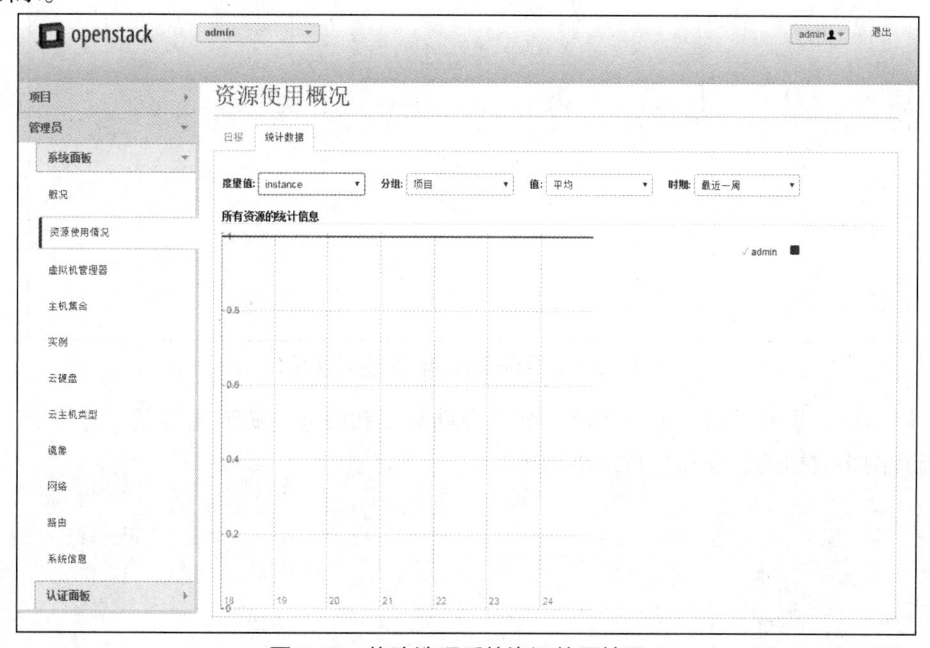

图 8-17　修改选项后的资源使用情况 1

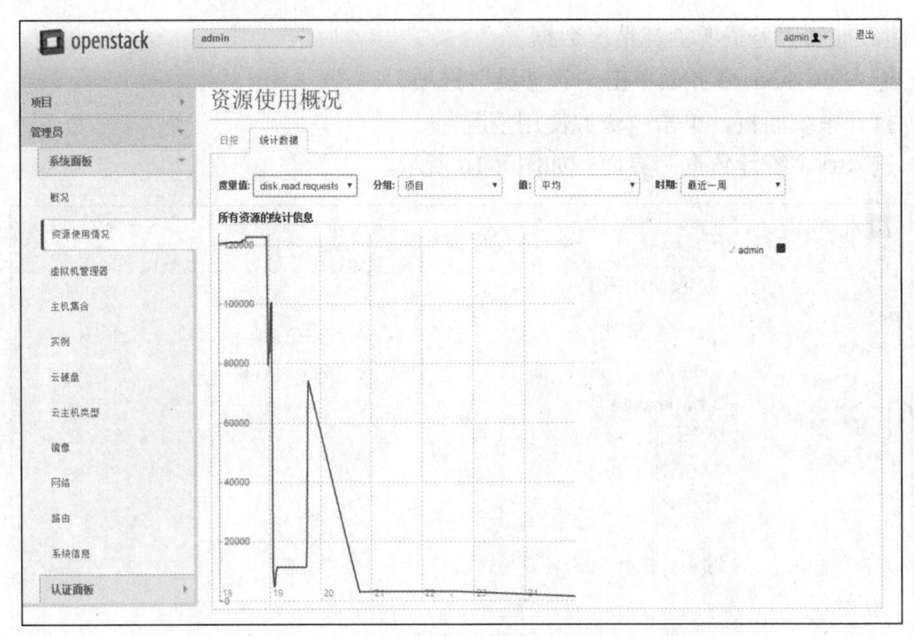

图 8-18　修改选项后的资源使用情况 2

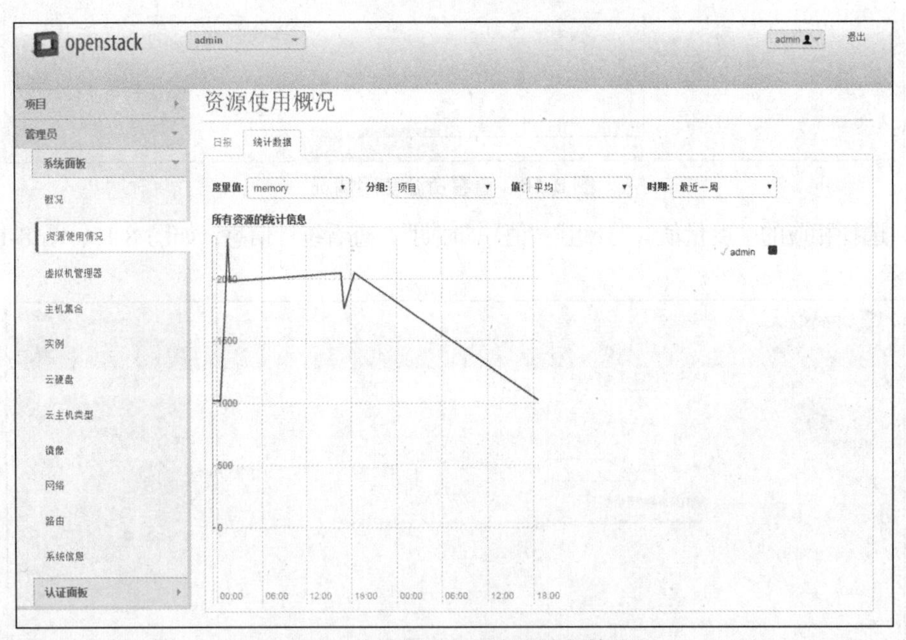

图 8-19　修改选项后的资源使用情况 3

可以用命令查看网络数据、实例数据、存储数据和服务资源消耗情况。

查看网络数据的命令和执行结果如下。

```
# ceilometer statistics -m network.incoming.bytes
```

Period	Period start	Period end	Max	Min	Avg	Sum	Count	Duration
0	2016-08-10 T06:53:36	2016-08-10 T06:53:36	13730.0	1996.0	10182.3807107	4011858.0	394	150001.0

查看实例数据的命令和执行结果如下。

```
# ceilometer statistics -m instance
```

Period	Period start	Period end	Max	Min	Avg	Sum	Count	Duration	Duration sta
0	2016-08-10 T06:53:36	2016-08-10 T06:53:36	1.0	1.0	1.0	405.0	405	150601.0	2016-08-10T 06:53:36

查看存储数据的命令和执行结果如下。

```
# ceilometer statistics -m disk.read.requests
```

Period	Period start	Period end	Max	Min	Avg	Sum	Count	Duration
0	2016-08 -10T06: 53:36	2016-08 -10T06: 53:36	3210.0	3010.0	3139.32323232	1243172.0	396	15060.0

查看服务资源消耗情况的命令和执行结果如下。

```
# ceilometer statistics -m memory
```

Period	Period start	Period end	Max	Min	Avg	Sum	Count	Duration
0	2016-08 -10T06: 53:36	2016-08 -10T06: 53:36	2048.0	1024	s	12288.0	9	12668.274

2. 数据库备份

辅助 Shell 脚本可做到对数据库的全量备份和增量备份，从而保证数据安全。

① 编写全量备份脚本。

MySQL 全量脚本内容见附录十一 mysql_full_bk.sh

② 开启 MySQL 数据库的 Binlog 功能，编写增量备份脚本。

```
# vi /etc/my.cnf
log-bin=mysql-bin
# servicemysqld restart
//MySQL 增量脚本内容见附录十二 mysql_hourly_bk.sh
```

③ 开启定时设置，编辑/etc/crontab 文件，添加以下内容。

```
#vi /etc/crontab
0 1-23/3 * * * (/bin/sh /root/mysql_hourly_bk.sh >> /opt/mysql/backup/ backup.log)
//在 1 至 23 时内每隔 3 小时执行全量备份命令
0 0 * * 0,1,3,6 (/bin/sh /root/mysql_full_bk.sh >> /opt/mysql/backup/ backup.log)
//每周一、三、六、七的 00:00 执行增量备份命令
# servicecrond restart
```

项目九 平台构建脚本解读

在平台构建和维护过程中，用户需要编写或使用第三方编写的大量脚本完成工作。因此，对脚本的理解程度也代表了用户对平台的理解程度和操作水平。平台构建脚本里涵盖了各个环节的参数、功能等，因此用户需要加强对平台构建脚本的理解。

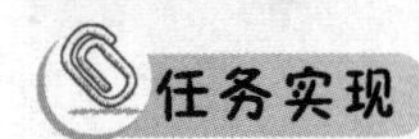

学习目标

- 详细了解和分析各个环节的脚本。
- 掌握脚本的含义和编写要点。

任务一 环境变量文件

任务实现

由于在 OpenStack 中需要修改和引用的变量文件相对较多，而且在后续的脚本安装过程中又需要进行重复地引用。因此，为了避免重复的工作，确保后续使用中能正确地调用所规定的值，可以编写环境变量的配置文件，使用规定名称命名，填入需要的数值，如下所示。

```
Mysql_Admin_Passwd=000000
Admin_Passwd=000000
Demo_User_Passwd=000000
Demo_DB_Passwd=000000
Contoller_Hostname=c1-controller
Controller_Mgmt_IPAddress=172.24.2.10
Gateway_Mgmt=172.24.2.1
Controller_External_IPAddress=172.24.3.10
```

上述文件中使用变量名“Mysql_Admin_Passwd”作为数据库密码的替代名称，后续文件中若涉及数据库密码可直接使用“$Mysql_Admin_Passwd”，即可调用到所赋予的 000000 值。这样有效地规范了数据库密码的使用，避免使用过程中出现的混乱和不一致。

任务二 网络模式

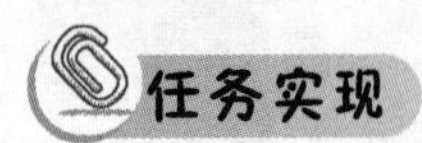

任务实现

使用节点安装脚本时，采用 OpenvSwitch 创建虚拟网卡和虚拟网桥，具体创建命令可见附录十三 ovs-network.txt。

在这段脚本中，使用 OpenvSwitch 的“ovs-vsctl add-br”命令分别创建 br-mgmt、br-eth0、br-eth1、br-prv、br-int 和 br-ex 网桥。其中，br-mgmt 为节点的管理网段通信网桥，br-eth0 为物理网口 eth0 网桥，br-eth1 为物理网口 eth1 网桥，br-prv 为实例通信私有网桥的网口，br-int 为实例创建连接网桥，br-ex 为实例对外通信网桥。

① 对这些网桥添加对等的虚拟端口。br-eth0 添加 eth0 和 br-eth0-br-mgmt 两个虚拟端口，分别作为与之对应端口的网桥名。br-mgmt 添加 br-mgmt-br-eth0 虚拟端口，作为与 eth0 连接的连接端口。

② 设置 eth0 网卡的连接端口。设置端口 br-eth0-br-mgmt、br-mgmt-br-eth0 类型为 patch，其中，patch 在 OpenStack 中可起到连接网桥的作用。

③ 设立 eth0 网卡的对等端口。设置端口 br-eth0-br-mgmt 和端口 br-mgmt-br-eth0，端口 br-mgmt-br-eth0 和端口 br-eth0-br-mgmt 同时为对等端口，设置两端互相连接通信。

④ 按照相同的步骤设置 eth1 网卡的连接端口，设置端口 br-prv-br-eth1、br-eth1-br-prv、br-eth1-br-ex、br-ex-br-eth1、br-eth1-br-int 和 br-int-br-eth1 的类型，同时设置对等端口，设置 br-prv-br-eth1 的对等端口为 br-eth1-br-prv，br-eth1-br-prv 的对等端口为 br-prv-br-eth1，br-eth1-br-ex 的对等端口为 br-ex-br-eth1，br-ex-br-eth1 的对等端口为 br-eth1-br-ex，br-eth1-br-int 的对等端口为 br-int-br-eth1，br-int-br-eth1 的对等端口为 br-eth1-br-int，设置完成后，设置外部连接的两个网桥 br-ex-br-eth1 和 br-eth1-br-ex 的端口类型为 Trunk，允许所有不同 Vlan ID 通过。

通过在节点输入命令，可以查询已配置的网络详情。具体查询结果可见附录十四 ovs-show.txt

```
Bridge "br-eth1"                    //OpenvSwitch eth1 网桥
    Port "br-eth1-br-ex"            //属于连接 br-ex 的端口
        trunks: [0]
        Interface "br-eth1-br-ex"
            type: patch
            options: {peer="br-ex-br-eth1"}
    Port "br-eth1-br-prv"           //属于连接 br-prv 的端口
        Interface "br-eth1-br-prv"
            type: patch
            options: {peer="br-prv-br-eth1"}
    Port "br-eth1"
        Interface "br-eth1"
            type: internal
    Port "eth1"
        Interface "eth1"
    Port "br-eth1-br-int"           //属于连接 br-eth 1 的端口
        Interface "br-eth1-br-int"
```

```
        type: patch
        options: {peer="br-int-br-eth1"}
```

任务三 节点安装脚本

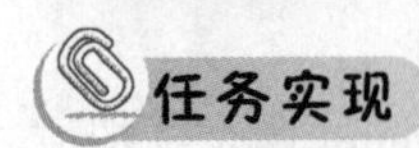

1. 控制节点

控制节点安装的软件均是控制端软件。首先，脚本执行软件安装的步骤，并且配置相关的系统基础服务，包括 IPTables、SELinux 和 IP 地址等，其中所需要的参数都是在基本环境配置文件中读取的，详情可见附录十五 environment.txt。

```
source $(pwd)/Xiandian_Pre.sh                          // 生效环境配置文件
iptables -F
iptables -X
iptables -Z
service iptables save                                  //清除防火墙规则保存配置
setenforce 0  2>&1                                     //修改 Selinux
```

默认配置信息修改完毕，统一创建数据库，修改所有服务的配置数据库连接，同时生成默认数据库表单，详情可见附录十六 mysql.txt。

```
service mysqld start                                   //启动数据库服务
create database IF NOT EXISTS keystone ;               //创建数据库
GRANT ALL PRIVILEGES ON keystone.* TO 'keystone'@'localhost' IDENTIFIED BY
'$Demo_DB_Passwd'
//配置每个服务的访问连接地址和赋予访问用户的权限
su -s /bin/sh -c "keystone-manage db_sync" keystone    //同步创建数据库表
```

至此，控制节点的安装全部完成。

2. 计算节点

计算节点主要配置 Nova 的 VNC 访问和 Neutron 提供的网络服务功能，部分脚本详情可见附录十七 compute.txt。

```
openstack-config --set /etc/nova/nova.conf DEFAULT auth_strategy keystone
                                                       //配置认证
openstack-config --set /etc/nova/nova.conf DEFAULT \
novncproxy_base_url http://$Controller_Mgmt_IPAddress:6080/vnc_auto.html
                                                       //配置 VNC 服务
openstack-config --set /etc/nova/nova.conf DEFAULT rpc_backend qpid
                                                       //配置监控和消息服务
openstack-config --set /etc/nova/nova.conf DEFAULT security_group_api neutron
//配置实例使用 Neutron 网络
```

至此，计算服务安装完毕。

附　　录

附录一　Xiandian_Pre.sh

参数配置脚本

```
######## Basic Environment ########

#          数据库用户密码            #
Mysql_Admin_Passwd=000000
#           管理员密码               #
Admin_Passwd=000000
#          演示用户密码              #
Demo_User_Passwd=000000
#           演示数据库密码           #
Demo_DB_Passwd=000000
###################################
#------------------------------------------------------------------------------
#  Controller   Node
#------------------------------------------------------------------------------
#  Controller Node Hostname
##         控制节点主机名           ##
Controller_Hostname=controller
#   Management Network of  Controller Node IP
##        控制节点管理地址          ##
Controller_Mgmt_IPAddress=172.24.2.10
##          管理网段网关            ##
Gateway_Mgmt=172.24.2.1
#   External Network of  Controller Node IP   #
##            外部地址              ##
Controller_External_IPAddress=172.24.3.10
#          Start Vlan ID              #
##        网络节点开始 Vlan ID     ##
Network_Start_Vlan_ID=43
```

```
#          End Vlan ID          #
##      网络节点结束 Vlan ID      ##
Network_End_Vlan_ID=46

#----------------------------------------------------------------------------
#  Compute   Node
#----------------------------------------------------------------------------
#  Compute Node Hostname
##        实例节点主机名          ##
Compute_Hostname=compute
#   Management Network of  Compute Node IP
##        实例节点管理地址        ##
Compute_Mgmt_IPAddress=172.24.2.20
#   External Network of  Compute Node IP
##       实例节点外部地址         ##
Compute_External_IPAddress=172.24.3.20
#   Cinder Disk Name eg (md126p3)
##    Cinder 存储磁盘分区名称      ##
Stroage_Cinder_Disk=sda3
#   Swift Disk Name eg (md126p4)
##    Swift 存储磁盘分区名称      ##
Stroage_Swift_Disk=sda2
```

附录二　Xiandian_Install_Controller_Node.sh

控制节点安装脚本

```
#!/bin/bash
source ./xiandian_pre.sh
#------------------------------ Install Packages ------------------------#
service NetworkManager stop 2>/dev/null
chkconfig NetworkManager off 2>/dev/null
setenforce 0  2>&1
yum upgrade -y
iptables -F
iptables -X
iptables -Z
service iptables save
```

```
sed -i "/REJECT/i\-A INPUT -m state --state NEW -m tcp -p tcp --dport 27017 \
-j ACCEPT" /etc/sysconfig/iptables
sed -i "/INPUT.*REJECT/i\-A INPUT -m state --state NEW -m tcp -p tcp --dport
5900:5909 \
-j ACCEPT" /etc/sysconfig/iptables
service iptables reload
sed -i 's/SELINUX=enforcing/SELINUX=permissive/g' /etc/selinux/config
cat <<EOF >> /etc/hosts
$Controller_Mgmt_IPAddress  $Controller_Hostname
$Compute_Mgmt_IPAddress     $Compute_Hostname
EOF
ssh-keygen
yum install -y ntp openstack-utils openssh-clients openstack-selinux openssh-
clients \
perl wget openssh-clients expect ntp openvswitch \
mysql mysql-server MySQL-python expect  \
openstack-keystone python-keystoneclient  \
openstack-utils openstack-selinux qpid-cpp-server \
openstack-glance python-glanceclient wget perl \
openstack-nova-api openstack-nova-cert openstack-nova-conductor \
openstack-nova-console openstack-nova-novncproxy openstack-nova-scheduler \
python-novaclient spice-html5 openstack-neutron openstack-neutron-ml2 \
python-neutronclient openstack-neutron-openvswitch \
memcached python-memcached mod_wsgi openstack-dashboard \
openstack-cinder openstack-utils openstack-selinux perl \
openstack-swift openstack-swift-proxy memcached \
python-swiftclient python-keystone-auth-token \
openstack-heat-api  openstack-heat-engine  openstack-heat-api-cfn  python-
openstackclient \
openstack-ceilometer-compute python-ceilometerclient python-pecan \
openstack-ceilometer-api openstack-ceilometer-collector \
openstack-ceilometer-notification openstack-ceilometer-central \
openstack-ceilometer-alarm python-ceilometerclient mongodb-server mongodb
for  hosts in $Controller_Hostname $Compute_Hostname; \
do ssh-copy-id -i /root/.ssh/id_rsa.pub $hosts  &&\
ssh $hosts "yum -y install openssh-clients";done
echo "server 127.127.1.0" >> /etc/ntp.conf
echo "fudge 127.127.1.0   stratum 10" >> /etc/ntp.conf
```

```
service ntpd restart
chkconfig ntpd on
#----------------------- Configure the Network ---------------------------#
service openvswitch start
chkconfig openvswitch on
sed  -i -e '/IPADDR/d' -e '/NETMASK/d' -e '/GATEWAY/d' -e '/DNS1/d' \
-e 's/ONBOOT=.*/ONBOOT=yes/g' -e 's/BOOTPROTO.*/BOOTPROTO=none/g' \
-e 's/NM_CONTROLLED=.*/NM_CONTROLLED=no/g' /etc/sysconfig/network-scripts/ ifcfg-
eth*
cat > /etc/sysconfig/network-scripts/ifcfg-br-mgmt  << EOF
DEVICE=br-mgmt
IPADDR=$Controller_Mgmt_IPAddress
NETMASK=255.255.255.0
BOOTPROTO=static
GATEWAY=$Gateway_Mgmt
ONBOOT=yes
USERCTL=no
EOF
cat > /etc/sysconfig/network-scripts/ifcfg-br-ex  << EOF
DEVICE=br-ex
IPADDR=$Controller_External_IPAddress
NETMASK=255.255.255.0
BOOTPROTO=static
ONBOOT=yes
USERCTL=no
EOF
ovs-vsctl add-br br-mgmt
ovs-vsctl add-br br-eth0
ovs-vsctl add-port br-eth0 eth0
ovs-vsctl add-port br-mgmt "br-mgmt-br-eth0"
ovs-vsctl add-port br-eth0 "br-eth0-br-mgmt"
ovs-vsctl set interface  "br-eth0-br-mgmt" type=patch
ovs-vsctl set interface  "br-mgmt-br-eth0" type=patch
ovs-vsctl set interface "br-eth0-br-mgmt" options:peer="br-mgmt-br-eth0"
ovs-vsctl set interface "br-mgmt-br-eth0" options:peer="br-eth0-br-mgmt"
ovs-vsctl add-br br-eth1
ovs-vsctl add-br br-prv
ovs-vsctl add-br br-int
```

```
ovs-vsctl add-br br-ex
ovs-vsctl add-port br-eth1 eth1
ovs-vsctl add-port br-eth1 "br-eth1-br-int"
ovs-vsctl add-port br-eth1 "br-eth1-br-prv"
ovs-vsctl add-port br-eth1 "br-eth1-br-ex"
ovs-vsctl add-port br-prv "br-prv-br-eth1"
ovs-vsctl add-port br-int "br-int-br-eth1"
ovs-vsctl add-port br-ex "br-ex-br-eth1"
ovs-vsctl set interface  "br-prv-br-eth1" type=patch
ovs-vsctl set interface  "br-eth1-br-prv" type=patch
ovs-vsctl set interface  "br-eth1-br-ex" type=patch
ovs-vsctl set interface  "br-ex-br-eth1" type=patch
ovs-vsctl set interface  "br-eth1-br-int" type=patch
ovs-vsctl set interface  "br-int-br-eth1" type=patch
ovs-vsctl set interface "br-prv-br-eth1" options:peer="br-eth1-br-prv"
ovs-vsctl set interface "br-eth1-br-prv" options:peer="br-prv-br-eth1"
ovs-vsctl set interface "br-eth1-br-ex" options:peer="br-ex-br-eth1"
ovs-vsctl set interface "br-ex-br-eth1" options:peer="br-eth1-br-ex"
ovs-vsctl set interface "br-eth1-br-int" options:peer="br-int-br-eth1"
ovs-vsctl set interface "br-int-br-eth1" options:peer="br-eth1-br-int"
ovs-vsctl set port "br-ex-br-eth1" trunks=0
ovs-vsctl set port "br-eth1-br-ex" trunks=0
service network restart
#----------------------- Configure the Environment ------------------------#
sed -i -e '/net.ipv4.ip_forward.*/d' -e '/net.ipv4.conf.all.rp_filter.*/d' \
-e '/net.ipv4.conf.default.rp_filter.*/d' /etc/sysctl.conf
cat >>/etc/sysctl.conf <<-EOF
net.ipv4.ip_forward=1
net.ipv4.conf.all.rp_filter=0
net.ipv4.conf.default.rp_filter=0
EOF
sysctl -p
sed -i '/^auth/s/yes/no/g' /etc/qpidd.conf
service qpidd start
chkconfig qpidd on
#------------------------ Configure the Databases -------------------------#
sed -i  "/^symbolic-links/a\default-storage-engine = \
innodb\ninnodb_file_per_table\ncollation-server = utf8_general_ci\ninit- connect = \
```

```
'SET NAMES utf8'\ncharacter-set-server = utf8" /etc/my.cnf
service mysqld start
chkconfig mysqld on
expect -c "
spawn /usr/bin/mysql_secure_installation
expect \"Enter current password for root (enter for none):\"
send \"\r\"
expect \"Set root password?\"
send \"y\r\"
expect \"New password:\"
send \"$Mysql_Admin_Passwd\r\"
expect \"Re-enter new password:\"
send \"$Mysql_Admin_Passwd\r\"
expect \"Remove anonymous users?\"
send \"y\r\"
expect \"Disallow root login remotely?\"
send \"n\r\"
expect \"Remove test database and access to it?\"
send \"y\r\"
expect \"Reload privilege tables now?\"
send \"y\r\"
expect eof
"
mysql -uroot -p$Mysql_Admin_Passwd -e "create database IF NOT EXISTS keystone ;"
mysql -uroot -p$Mysql_Admin_Passwd -e "GRANT ALL PRIVILEGES ON keystone.* \
TO 'keystone'@'localhost' IDENTIFIED BY '$Demo_DB_Passwd' ;"
mysql -uroot -p$Mysql_Admin_Passwd -e "GRANT ALL PRIVILEGES ON keystone.* \
TO 'keystone'@'%' IDENTIFIED BY '$Demo_DB_Passwd' ;"
mysql -uroot -p$Mysql_Admin_Passwd -e "create database IF NOT EXISTS glance ;"
mysql -uroot -p$Mysql_Admin_Passwd -e "GRANT ALL PRIVILEGES ON glance.* \
TO 'glance'@'localhost' IDENTIFIED BY '$Demo_DB_Passwd' ;"
mysql -uroot -p$Mysql_Admin_Passwd -e "GRANT ALL PRIVILEGES ON glance.* \
TO 'glance'@'%' IDENTIFIED BY '$Demo_DB_Passwd' ;"
mysql -uroot -p$Mysql_Admin_Passwd -e "create database IF NOT EXISTS nova ;"
mysql -uroot -p$Mysql_Admin_Passwd -e "GRANT ALL PRIVILEGES ON nova.* \
TO 'nova'@'localhost' IDENTIFIED BY '$Demo_DB_Passwd' ;"
mysql -uroot -p$Mysql_Admin_Passwd -e "GRANT ALL PRIVILEGES ON nova.* \
TO 'nova'@'%' IDENTIFIED BY '$Demo_DB_Passwd' ;"
```

```
mysql -uroot -p$Mysql_Admin_Passwd -e "create database IF NOT EXISTS neutron ;"
mysql -uroot -p$Mysql_Admin_Passwd -e "GRANT ALL PRIVILEGES ON neutron.* \
TO 'neutron'@'localhost' IDENTIFIED BY '$Demo_DB_Passwd' ;"
mysql -uroot -p$Mysql_Admin_Passwd -e "GRANT ALL PRIVILEGES ON neutron.* \
TO 'neutron'@'%' IDENTIFIED BY '$Demo_DB_Passwd' ;"
mysql -uroot -p$Mysql_Admin_Passwd -e "create database IF NOT EXISTS cinder ;"
mysql -uroot -p$Mysql_Admin_Passwd -e "GRANT ALL PRIVILEGES ON cinder.* \
TO 'cinder'@'localhost' IDENTIFIED BY '$Demo_DB_Passwd' ;"
mysql -uroot -p$Mysql_Admin_Passwd -e "GRANT ALL PRIVILEGES ON cinder.* \
TO 'cinder'@'%' IDENTIFIED BY '$Demo_DB_Passwd' ;"
mysql -uroot -p$Mysql_Admin_Passwd -e "create database IF NOT EXISTS heat ;"
mysql -uroot -p$Mysql_Admin_Passwd -e "GRANT ALL PRIVILEGES ON heat.* \
TO 'heat'@'localhost' IDENTIFIED BY '$Demo_DB_Passwd' ;"
mysql -uroot -p$Mysql_Admin_Passwd -e "GRANT ALL PRIVILEGES ON heat.* \
TO 'heat'@'%' IDENTIFIED BY '$Demo_DB_Passwd' ;"
sed -i "s/bind_ip.*/bind_ip = 127.0.0.1,$Controller_Mgmt_IPAddress/g" /etc/
mongodb.conf
echo "smallfiles = true">>/etc/mongodb.conf
service mongod restart
chkconfig mongod on
echo -e "\033[31m    Please waiting......         \033[0m "
sleep 90
mongo $Controller_Mgmt_IPAddress/ceilometer --eval "db.addUser({user: 'ceilometer', \
pwd: '$Demo_DB_Passwd', roles: [ 'readWrite', 'dbAdmin' ]})"
while [ $? -ne 0 ]
do
sleep 60
echo "Please waiting......"
mongo $Controller_Mgmt_IPAddress/ceilometer --eval "db.addUser({user: 'ceilometer', \
pwd: '$Demo_DB_Passwd', roles: [ 'readWrite', 'dbAdmin' ]})"
done

#-------------------- Configure the database and tables   ------------------#
# Configure databases connectios #
openstack-config --set /etc/keystone/keystone.conf database \
connection mysql://keystone:$Demo_DB_Passwd@$Controller_Mgmt_IPAddress/keystone
openstack-config --set /etc/glance/glance-api.conf database \
connection mysql://glance:$Demo_DB_Passwd@$Controller_Mgmt_IPAddress/glance
```

```
openstack-config --set /etc/glance/glance-registry.conf database \
connection mysql://glance:$Demo_DB_Passwd@$Controller_Mgmt_IPAddress/glance
openstack-config --set /etc/nova/nova.conf database \
connection mysql://nova:$Demo_DB_Passwd@$Controller_Mgmt_IPAddress/nova
openstack-config --set /etc/neutron/neutron.conf database \
connection mysql://neutron:$Demo_DB_Passwd@$Controller_Mgmt_IPAddress/neutron
openstack-config --set /etc/cinder/cinder.conf database \
connection mysql://cinder:$Demo_DB_Passwd@$Controller_Mgmt_IPAddress/cinder
openstack-config --set /etc/heat/heat.conf database \
connection mysql://heat:$Demo_DB_Passwd@$Controller_Mgmt_IPAddress/heat
openstack-config --set /etc/ceilometer/ceilometer.conf database connection \
mongodb://ceilometer:$Demo_DB_Passwd@$Controller_Mgmt_IPAddress:27017/ceilometer
# Update Services Tables #
su -s /bin/sh -c "keystone-manage db_sync" keystone
su -s /bin/sh -c "glance-manage db_sync" glance
su -s /bin/sh -c "nova-manage db sync" nova
su -s /bin/sh -c "cinder-manage db sync" cinder
su -s /bin/sh -c "heat-manage db_sync" heat
#--------------------- Configure the Keystone ----------------------------#
ADMIN_TOKEN=$(openssl rand -hex 10)
openstack-config --set /etc/keystone/keystone.conf DEFAULT admin_token $ADMIN_
TOKEN
keystone-manage pki_setup --keystone-user keystone --keystone-group keystone
chown -R keystone:keystone /etc/keystone/* /var/log/keystone/*
chmod -R o-rwx /etc/keystone/ssl
chkconfig openstack-keystone on
service openstack-keystone start
touch /var/log/keystone/keystone-tokenflush.log
chmod 777 /var/log/keystone/keystone-tokenflush.log
(crontab -l -u keystone 2>/var/log/keystone/keystone-tokenflush.log | grep -q
token_flush) || \
echo '@hourly /usr/bin/keystone-manage token_flush >/var/log/keystone/ keystone-
tokenflush.log \
2>&1' >> /var/spool/cron/keystone
export OS_SERVICE_TOKEN=$ADMIN_TOKEN
export OS_SERVICE_ENDPOINT=http://$Controller_Mgmt_IPAddress:35357/v2.0
#------------------------ Create Keystone Users ------------------------#
keystone user-create --name=admin --pass=$Admin_Passwd
```

```
keystone role-create --name=admin
keystone tenant-create --name=admin --description="Admin Tenant"
keystone user-role-add --user=admin --tenant=admin --role=admin
keystone user-role-add --user=admin --role=_member_ --tenant=admin
keystone user-create --name=demo --pass=$Demo_User_Passwd
keystone tenant-create --name=demo --description="Demo Tenant"
keystone user-role-add --user=demo --role=_member_ --tenant=demo
keystone tenant-create --name=service --description="Service Tenant"
keystone service-create --name=keystone --type=identity --description= "OpenStack
Identity"
keystone endpoint-create --service-id=$(keystone service-list | awk '/ identity
/ {print $2}') \
--publicurl=http://$Controller_External_IPAddress:5000/v2.0 \
--internalurl=http://$Controller_Mgmt_IPAddress:5000/v2.0 \
--adminurl=http://$Controller_Mgmt_IPAddress:35357/v2.0
unset OS_SERVICE_TOKEN OS_SERVICE_ENDPOINT
cat > /etc/keystone/admin-openrc.sh <<-EOF
export OS_USERNAME=admin
export OS_PASSWORD=$Admin_Passwd
export OS_TENANT_NAME=admin
export OS_AUTH_URL=http://$Controller_Mgmt_IPAddress:35357/v2.0
EOF
cat > /etc/keystone/demo-openrc.sh <<-EOF
export OS_USERNAME=demo
export OS_PASSWORD=$Demo_User_Passwd
export OS_TENANT_NAME=demo
export OS_AUTH_URL=http://$Controller_Mgmt_IPAddress:35357/v2.0
EOF
source /etc/keystone/admin-openrc.sh
#-------------------------- Create for Glance ---------------------------#
keystone user-create --name=glance --pass=$Demo_User_Passwd
keystone user-role-add --user=glance --tenant=service --role=admin
keystone service-create --name=glance --type=image --description=" OpenStack Image
Service"
keystone endpoint-create --service-id=$(keystone service-list | awk '/ image /
{print $2}') \
--publicurl=http://$Controller_External_IPAddress:9292 \
--internalurl=http://$Controller_Mgmt_IPAddress:9292 \
```

```
--adminurl=http://$Controller_Mgmt_IPAddress:9292
#---------------------------- Create for Nova ----------------------------#
keystone user-create --name=nova --pass=$Demo_User_Passwd
keystone user-role-add --user=nova --tenant=service --role=admin
keystone service-create --name=nova --type=compute --description="OpenStack Compute"
keystone endpoint-create --service-id=$(keystone service-list | awk '/ compute
/ {print $2}') \
--publicurl=http://$Controller_External_IPAddress:8774/v2/%\(tenant_id\)s \
--internalurl=http://$Controller_Mgmt_IPAddress:8774/v2/%\(tenant_id\)s \
--adminurl=http://$Controller_Mgmt_IPAddress:8774/v2/%\(tenant_id\)s
#---------------------------- Create for Neutron --------------------------#
keystone user-create --name neutron --pass $Demo_User_Passwd
keystone user-role-add --user neutron --tenant service --role admin
keystone service-create --name neutron --type network --description "OpenStack
Networking"
keystone endpoint-create --service-id $(keystone service-list | awk '/ network
/ {print $2}') \
--publicurl http://$Controller_External_IPAddress:9696 \
--adminurl http://$Controller_Mgmt_IPAddress:9696 \
--internalurl http://$Controller_Mgmt_IPAddress:9696
#------------------------- Create for Cinder ------------------------------#
keystone user-create --name=cinder --pass=$Demo_User_Passwd
keystone user-role-add --user=cinder --tenant=service --role=admin
keystone  service-create  --name=cinder  --type=volume  --description="OpenStack
Block Storage"
keystone endpoint-create --service-id=$(keystone service-list | awk '/ volume
/ {print $2}') \
--publicurl=http://$Controller_External_IPAddress:8776/v1/%\(tenant_id\)s \
--internalurl=http://$Controller_Mgmt_IPAddress:8776/v1/%\(tenant_id\)s \
--adminurl=http://$Controller_Mgmt_IPAddress:8776/v1/%\(tenant_id\)s
keystone service-create --name=cinderv2 --type=volumev2 \
--description="OpenStack Block Storage v2"
keystone endpoint-create --service-id=$(keystone service-list | awk '/ volumev2
/ {print $2}') \
--publicurl=http://$Controller_External_IPAddress:8776/v2/%\(tenant_id\)s \
--internalurl=http://$Controller_Mgmt_IPAddress:8776/v2/%\(tenant_id\)s \
--adminurl=http://$Controller_Mgmt_IPAddress:8776/v2/%\(tenant_id\)s
#--------------------------- Create for Swift ----------------------------#
```

```
keystone user-create --name=swift --pass=$Demo_User_Passwd
keystone role-create --name=SwiftOperator
keystone user-role-add --user=swift --tenant=service --role=admin
keystone user-role-add --user=swift --tenant=service --role=SwiftOperator
keystone service-create --name=swift --type=object-store \
--description="OpenStack Object Storage"
keystone endpoint-create --service-id=$(keystone service-list | awk '/ object-
store / {print $2}') \
--publicurl="http://$Controller_External_IPAddress:8080/v2/AUTH_%(tenant_id)s" \
--internalurl="http://$Controller_Mgmt_IPAddress:8080/v2/AUTH_%(tenant_id)s" \
--adminurl="http://$Controller_Mgmt_IPAddress:8080/v2/AUTH_%(tenant_id)s"
#---------------------------- Create for heat -----------------------------#
keystone user-create --name=heat --pass=$Demo_User_Passwd
keystone user-role-add --user=heat --tenant=service --role=admin
keystone service-create --name=heat --type=orchestration --description=
"Orchestration"
keystone endpoint-create --service-id=$(keystone service-list | awk '/
orchestration / {print $2}') \
--publicurl=http://$Controller_External_IPAddress:8004/v1/%\(tenant_id\)s \
--internalurl=http://$Controller_Mgmt_IPAddress:8004/v1/%\(tenant_id\)s \
--adminurl=http://$Controller_Mgmt_IPAddress:8004/v1/%\(tenant_id\)s
keystone service-create --name=heat-cfn --type=cloudformation \
--description="Orchestration CloudFormation"
keystone endpoint-create --service-id=$(keystone service-list | awk '/
cloudformation / {print $2}') \
--publicurl=http://$Controller_External_IPAddress:8000/v1 \
--internalurl=http://$Controller_Mgmt_IPAddress:8000/v1 \
--adminurl=http://$Controller_Mgmt_IPAddress:8000/v1
#------------------------- Create for Ceilometer -------------------------#
keystone user-create --name=ceilometer --pass=$Demo_User_Passwd
keystone user-role-add --user=ceilometer --tenant=service --role=admin
keystone service-create --name=ceilometer --type=metering --description=
"Telemetry"
keystone endpoint-create --service-id=$(keystone service-list | awk '/ metering
/ {print $2}') \
--publicurl=http://$Controller_External_IPAddress:8777 \
--internalurl=http://$Controller_Mgmt_IPAddress:8777 \
--adminurl=http://$Controller_Mgmt_IPAddress:8777
```

```
keystone role-create --name=ResellerAdmin
keystone user-role-add --tenant service --user ceilometer \
--role $(keystone role-list | awk '/ ResellerAdmin / {print $2}')
###################### Configuration Services ########################
#----------------------- Configuration Glance -------------------------#
perl -p -i -e "s/^(if _fastmath is not None .*:)/#\1/" \
/usr/lib64/python2.6/site-packages/Crypto/Util/number.py
perl -p -i -e "s/^(\s*_warn.*Not using mpz_powm_sec.*)/#\1/" \
/usr/lib64/python2.6/site-packages/Crypto/Util/number.py
openstack-config --set /etc/glance/glance-api.conf keystone_authtoken \
auth_uri http://$Controller_Mgmt_IPAddress:5000
openstack-config --set /etc/glance/glance-api.conf keystone_authtoken \
auth_host $Controller_Mgmt_IPAddress
openstack-config --set /etc/glance/glance-api.conf keystone_authtoken auth_
port 35357
openstack-config --set /etc/glance/glance-api.conf keystone_authtoken auth_
protocol http
openstack-config --set /etc/glance/glance-api.conf keystone_authtoken \
admin_tenant_name service
openstack-config --set /etc/glance/glance-api.conf keystone_authtoken admin_
user glance
openstack-config --set /etc/glance/glance-api.conf keystone_authtoken \
admin_password $Demo_User_Passwd
openstack-config --set /etc/glance/glance-api.conf paste_deploy flavor keystone
openstack-config --set /etc/glance/glance-api.conf DEFAULT notification_ driver
messaging
openstack-config --set /etc/glance/glance-registry.conf keystone_authtoken \
auth_uri http://$Controller_Mgmt_IPAddress:5000
openstack-config --set /etc/glance/glance-registry.conf keystone_authtoken \
auth_host $Controller_Mgmt_IPAddress
openstack-config --set /etc/glance/glance-registry.conf keystone_authtoken \
auth_port 35357
openstack-config --set /etc/glance/glance-registry.conf keystone_authtoken \
auth_protocol http
openstack-config --set /etc/glance/glance-registry.conf keystone_authtoken \
admin_tenant_name service
openstack-config --set /etc/glance/glance-registry.conf keystone_authtoken
admin_user glance
```

```
openstack-config --set /etc/glance/glance-registry.conf keystone_authtoken \
admin_password $Demo_User_Passwd
openstack-config --set /etc/glance/glance-registry.conf paste_deploy flavor
keystone
openstack-config --set /etc/glance/glance-api.conf DEFAULT rpc_backend qpid
#------------------------ Configuration Nova -------------------------#
openstack-config --set /etc/nova/nova.conf DEFAULT rpc_backend qpid
openstack-config --set /etc/nova/nova.conf DEFAULT qpid_hostname $Controller_
Mgmt_IPAddress
openstack-config --set /etc/nova/nova.conf DEFAULT my_ip $Controller_ Mgmt_
IPAddress
openstack-config --set /etc/nova/nova.conf DEFAULT vncserver_listen $Controller_
Mgmt_IPAddress
openstack-config --set /etc/nova/nova.conf DEFAULT \
vncserver_proxyclient_address $Controller_Mgmt_IPAddress
openstack-config --set /etc/nova/nova.conf DEFAULT auth_strategy keystone
openstack-config --set /etc/nova/nova.conf keystone_authtoken \
auth_uri http://$Controller_Mgmt_IPAddress:5000
openstack-config --set /etc/nova/nova.conf keystone_authtoken auth_host
$Controller_Mgmt_IPAddress
openstack-config --set /etc/nova/nova.conf keystone_authtoken auth_protocol http
openstack-config --set /etc/nova/nova.conf keystone_authtoken auth_port 35357
openstack-config --set /etc/nova/nova.conf keystone_authtoken admin_user nova
openstack-config --set /etc/nova/nova.conf keystone_authtoken admin_tenant_name
service
openstack-config --set /etc/nova/nova.conf keystone_authtoken admin_password
$Demo_User_Passwd
#----------------------- Configuration Neutron -------------------------#
openstack-config --set /etc/neutron/metadata_agent.ini DEFAULT \
auth_url http://$Controller_Hostname:5000/v2.0
openstack-config --set /etc/neutron/metadata_agent.ini DEFAULT auth_ region
regionOne
openstack-config --set /etc/neutron/metadata_agent.ini DEFAULT admin_ tenant_
name service
openstack-config --set /etc/neutron/metadata_agent.ini DEFAULT admin_user
neutron
openstack-config --set /etc/neutron/metadata_agent.ini DEFAULT admin_ password
$Demo_User_Passwd
```

```
openstack-config --set /etc/neutron/metadata_agent.ini DEFAULT \
nova_metadata_ip $Controller_Mgmt_IPAddress
openstack-config --set /etc/neutron/metadata_agent.ini DEFAULT metadata_ proxy_
shared_secret 000000
openstack-config --set /etc/neutron/neutron.conf DEFAULT auth_strategy keystone
openstack-config --set /etc/neutron/neutron.conf DEFAULT \
rpc_backend neutron.openstack.common.rpc.impl_qpid
openstack-config --set /etc/neutron/neutron.conf DEFAULT qpid_hostname $Controller_
Mgmt_IPAddress
openstack-config --set /etc/neutron/neutron.conf DEFAULT core_plugin ml2
openstack-config --set /etc/neutron/neutron.conf DEFAULT service_plugins router
openstack-config --set /etc/neutron/neutron.conf DEFAULT control_exchange neutron
openstack-config --set /etc/neutron/neutron.conf DEFAULT \
notification_driver neutron.openstack.common.notifier.rpc_notifier
openstack-config --set /etc/neutron/neutron.conf keystone_authtoken \
auth_uri http://$Controller_Mgmt_IPAddress:5000
openstack-config --set /etc/neutron/neutron.conf keystone_authtoken auth_ host
$Controller_Mgmt_IPAddress
openstack-config --set /etc/neutron/neutron.conf keystone_authtoken auth_
protocol http
openstack-config --set /etc/neutron/neutron.conf keystone_authtoken auth_ port
35357
openstack-config --set /etc/neutron/neutron.conf keystone_authtoken admin_ tenant_
name service
openstack-config --set /etc/neutron/neutron.conf keystone_authtoken admin_ user
neutron
openstack-config --set /etc/neutron/neutron.conf keystone_authtoken admin_ password
$Demo_User_Passwd
openstack-config --set /etc/neutron/dhcp_agent.ini DEFAULT \
interface_driver neutron.agent.linux.interface.OVSInterfaceDriver
openstack-config --set /etc/neutron/dhcp_agent.ini DEFAULT dhcp_driver neutron.
agent.linux.dhcp.Dnsmasq
openstack-config --set /etc/neutron/dhcp_agent.ini DEFAULT use_namespaces True
openstack-config --set /etc/neutron/l3_agent.ini DEFAULT \
interface_driver neutron.agent.linux.interface.OVSInterfaceDriver
openstack-config --set /etc/neutron/l3_agent.ini DEFAULT use_namespaces True
openstack-config --set /etc/neutron/l3_agent.ini DEFAULT external_network_ bridge
br-ex
```

```
openstack-config --set /etc/neutron/l3_agent.ini DEFAULT metadata_port 8775
openstack-config --set /etc/neutron/l3_agent.ini DEFAULT admin_password $Demo_
User_Passwd
openstack-config --set /etc/neutron/l3_agent.ini DEFAULT admin_tenant_ name
service
openstack-config --set /etc/neutron/l3_agent.ini DEFAULT \
auth_url http://$Controller_Mgmt_IPAddress:35357/v2.0
openstack-config --set /etc/neutron/l3_agent.ini DEFAULT verbose true
openstack-config --set /etc/neutron/l3_agent.ini DEFAULT metadata_ip 169.254.
169.254
openstack-config --set /etc/neutron/l3_agent.ini DEFAULT admin_user neutron
openstack-config --set /etc/nova/nova.conf DEFAULT network_api_class nova.
network.neutronv2.api.API
openstack-config --set /etc/nova/nova.conf DEFAULT neutron_url http:// $Controller_
Mgmt_IPAddress:9696
openstack-config --set /etc/nova/nova.conf DEFAULT neutron_auth_strategy
keystone
openstack-config --set /etc/nova/nova.conf DEFAULT neutron_admin_tenant_ name
service
openstack-config --set /etc/nova/nova.conf DEFAULT neutron_admin_username
neutron
openstack-config --set /etc/nova/nova.conf DEFAULT neutron_admin_password
$Demo_User_Passwd
openstack-config --set /etc/nova/nova.conf DEFAULT \
neutron_admin_auth_url http://$Controller_Mgmt_IPAddress:35357/v2.0
openstack-config --set /etc/nova/nova.conf DEFAULT \
linuxnet_interface_driver nova.network.linux_net.LinuxOVSInterfaceDriver
openstack-config --set /etc/nova/nova.conf DEFAULT firewall_driver nova.virt.
firewall.NoopFirewallDriver
openstack-config --set /etc/nova/nova.conf DEFAULT security_group_api neutron
openstack-config --set /etc/nova/nova.conf DEFAULT vif_plugging_is_fatal False
openstack-config --set /etc/nova/nova.conf DEFAULT vif_plugging_timeout 10
openstack-config --set /etc/nova/nova.conf DEFAULT service_neutron_ metadata_
proxy true
openstack-config --set /etc/nova/nova.conf DEFAULT neutron_metadata_ proxy_
shared_secret 000000
openstack-config --set /etc/neutron/neutron.conf DEFAULT notify_nova_on_ port_
status_changes True
```

```
openstack-config --set /etc/neutron/neutron.conf DEFAULT notify_nova_on_ port_
data_changes True
openstack-config --set /etc/neutron/neutron.conf DEFAULT \
nova_url http://$Controller_Mgmt_IPAddress:8774/v2
openstack-config --set /etc/neutron/neutron.conf DEFAULT nova_admin_ username
nova
openstack-config --set /etc/neutron/neutron.conf DEFAULT \
nova_admin_tenant_id $(keystone tenant-list | awk '/ service / { print $2 }')
openstack-config --set /etc/neutron/neutron.conf DEFAULT nova_admin_ password
$Demo_User_Passwd
openstack-config --set /etc/neutron/neutron.conf DEFAULT \
nova_admin_auth_url http://$Controller_Mgmt_IPAddress:35357/v2.0
openstack-config --set /etc/neutron/plugins/ml2/ml2_conf.ini ml2 type_drivers vlan
openstack-config  --set  /etc/neutron/plugins/ml2/ml2_conf.ini  ml2  tenant_
network_types  vlan
openstack-config  --set  /etc/neutron/plugins/ml2/ml2_conf.ini  ml2  mechanism_
drivers  openvswitch
openstack-config --set /etc/neutron/plugins/ml2/ml2_conf.ini ml2_type_ vlan \
network_vlan_ranges physnet1:$Network_Start_Vlan_ID:$Network_End_Vlan_ID
openstack-config  --set  /etc/neutron/plugins/ml2/ml2_conf.ini  ovs  enable_
tunneling False
openstack-config --set /etc/neutron/plugins/ml2/ml2_conf.ini ovs integration_ bridge
br-int
openstack-config  --set  /etc/neutron/plugins/ml2/ml2_conf.ini  ovs  bridge_
mappings physnet1:br-prv
openstack-config --set /etc/neutron/plugins/ml2/ml2_conf.ini securitygroup  \
firewall_driver neutron.agent.linux.iptables_firewall.OVSHybridIptables FirewallDriver
openstack-config --set /etc/neutron/plugins/ml2/ml2_conf.ini securitygroup  \
enable_security_group True
ln -s plugins/ml2/ml2_conf.ini  /etc/neutron/plugin.ini
cp /etc/init.d/neutron-openvswitch-agent  /etc/init.d/neutron-openvswitch- agent.
orig
sed -i 's,plugins/openvswitch/ovs_neutron_plugin.ini,plugin.ini,g' /etc/ init.d/
neutron-openvswitch-agent
#------------------------ Configuration Cinder ---------------------------#
sed -i -e "s/^ALLOWED_HOSTS.*/ALLOWED_HOSTS = ['*','localhost']/g" \
-e "s/^OPENSTACK_HOST.*/OPENSTACK_HOST = \"$Controller_Mgmt_IPAddress\"/g" \
-e "s/^TIME_ZONE.*/TIME_ZONE = \"UTC\"/g" -e "/^CACHES = {/,/^}/s/'LOCAT.*//g" \
```

```
-e "/^CACHES = {/,/^}/s/^        'BACKEND'.*\
/   'BACKEND' : 'django.core.cache.backends.memcached.MemcachedCache',\n   \
'LOCATION' : '$Controller_Mgmt_IPAddress:11211',/g" /etc/openstack-dashboard/ local_
settings
perl -p -i -e "s/^(if _fastmath is not None .*:)/#\1/" /usr/lib64/python2.6/
site-packages/Crypto/Util/number.py
perl -p -i -e "s/^(\s*_warn.*Not using mpz_powm_sec.*)/#\1/" \
/usr/lib64/python2.6/site-packages/Crypto/Util/number.py
openstack-config --set /etc/cinder/cinder.conf DEFAULT auth_strategy keystone
openstack-config --set /etc/cinder/cinder.conf keystone_authtoken \
auth_uri http://$Controller_Mgmt_IPAddress:5000
openstack-config --set /etc/cinder/cinder.conf keystone_authtoken auth_ host
$Controller_Mgmt_IPAddress
openstack-config --set /etc/cinder/cinder.conf keystone_authtoken auth_protocol http
openstack-config --set /etc/cinder/cinder.conf keystone_authtoken auth_ port
35357
openstack-config --set /etc/cinder/cinder.conf keystone_authtoken admin_ user
cinder
openstack-config --set /etc/cinder/cinder.conf keystone_authtoken admin_ tenant_
name service
openstack-config  --set  /etc/cinder/cinder.conf  keystone_authtoken  admin_
password $Demo_User_Passwd
openstack-config --set /etc/cinder/cinder.conf DEFAULT \
rpc_backend cinder.openstack.common.rpc.impl_qpid
openstack-config   --set   /etc/cinder/cinder.conf   DEFAULT   qpid_hostname
$Controller_Mgmt_IPAddress
openstack-config --set /etc/cinder/cinder.conf DEFAULT control_exchange cinder
openstack-config --set /etc/cinder/cinder.conf DEFAULT \
notification_driver cinder.openstack.common.notifier.rpc_notifier
openstack-config --set /etc/cinder/cinder.conf DEFAULT iscsi_helper tgtadm
openstack-config --set /usr/share/cinder/cinder-dist.conf DEFAULT iscsi_ helper
tgtadm
#----------------------- Configuration Swift --------------------------#
mkdir -p /etc/swift/
openstack-config --set /etc/swift/swift.conf swift-hash \
swift_hash_path_suffix 'od -t x8 -N 8 -A n </dev/random'
openstack-config --set /etc/swift/proxy-server.conf DEFAULT bind_port 8080
openstack-config --set /etc/swift/proxy-server.conf DEFAULT workers 8
```

```
openstack-config --set /etc/swift/proxy-server.conf DEFAULT user swift
openstack-config --set /etc/swift/proxy-server.conf DEFAULT log_name proxy
openstack-config --set /etc/swift/proxy-server.conf DEFAULT log_level DEBUG
openstack-config --set /etc/swift/proxy-server.conf DEFAULT log_facility
LOG_LOCAL0
openstack-config --set /etc/swift/proxy-server.conf pipeline:main \
pipeline healthcheck\ cache\ authtoken\ keystone\ proxy-server
openstack-config --set /etc/swift/proxy-server.conf app:proxy-server use
egg:swift#proxy
openstack-config --set /etc/swift/proxy-server.conf app:proxy-server allow_
account_management true
openstack-config --set /etc/swift/proxy-server.conf app:proxy-server account_
autocreate true
openstack-config --set /etc/swift/proxy-server.conf filter:cache use egg:
swift#memcache
openstack-config --set /etc/swift/proxy-server.conf filter:cache memcache_
servers 127.0.0.1:11211
openstack-config --set /etc/swift/proxy-server.conf filter:catch_errors use
egg:swift#catch_errors
openstack-config --set /etc/swift/proxy-server.conf filter:healthcheck use
egg:swift#healthcheck
openstack-config --set /etc/swift/proxy-server.conf filter:keystone use
egg:swift#keystoneauth
openstack-config --set /etc/swift/proxy-server.conf filter:keystone \
operator_roles admin,\ SwiftOperator,\ _member_
openstack-config --set /etc/swift/proxy-server.conf filter:keystone is_ admin
true
openstack-config --set /etc/swift/proxy-server.conf filter:keystone cache
swift.cache
openstack-config --set /etc/swift/proxy-server.conf filter:authtoken \
paste.filter_factory keystoneclient.middleware.auth_token:filter_factory
openstack-config --set /etc/swift/proxy-server.conf filter:authtoken signing_
dir /tmp/keystone-signing-swift
openstack-config --set /etc/swift/proxy-server.conf filter:authtoken auth_
protocol http
openstack-config --set /etc/swift/proxy-server.conf filter:authtoken auth_ host
$Controller_Mgmt_IPAddress
openstack-config --set /etc/swift/proxy-server.conf filter:authtoken auth_ port
```

```
35357
openstack-config --set /etc/swift/proxy-server.conf filter:authtoken admin_
tenant_name service
openstack-config --set /etc/swift/proxy-server.conf filter:authtoken admin_
user swift
openstack-config --set /etc/swift/proxy-server.conf filter:authtoken \
admin_password $Demo_User_Passwd
pushd /etc/swift
swift-ring-builder account.builder create 18 1 1
swift-ring-builder container.builder create 18 1 1
swift-ring-builder object.builder create 18 1 1
swift-ring-builder account.builder add z1-$Compute_Mgmt_IPAddress:6002/ $Stroage_
Swift_Disk 100
swift-ring-builder  container.builder  add  z1-$Compute_Mgmt_IPAddress:6001/
$Stroage_ Swift_Disk 100
swift-ring-builder object.builder add z1-$Compute_Mgmt_IPAddress:6000/ $Stroage_
Swift_Disk 100
swift-ring-builder account.builder rebalance
swift-ring-builder container.builder rebalance
swift-ring-builder object.builder rebalance
popd
echo "local0.* /var/log/swift/proxy.log">>/etc/rsyslog.d/10-swift.conf
mkdir -p /var/log/swift
chown -R swift:swift /var/log/swift
chown -R swift:swift /etc/swift
#-------------------------- Configuration Heat --------------------------#
openstack-config --set /etc/heat/heat.conf DEFAULT qpid_hostname $Controller_Mgmt_
IPAddress
openstack-config --set /etc/heat/heat.conf keystone_authtoken auth_host $Controller_
Mgmt_IPAddress
openstack-config --set /etc/heat/heat.conf keystone_authtoken auth_port 35357
openstack-config --set /etc/heat/heat.conf keystone_authtoken auth_protocol http
openstack-config --set /etc/heat/heat.conf keystone_authtoken \
auth_uri http://$Controller_Mgmt_IPAddress:5000/v2.0
openstack-config --set /etc/heat/heat.conf keystone_authtoken admin_tenant_ name
service
openstack-config --set /etc/heat/heat.conf keystone_authtoken admin_user heat
openstack-config  --set  /etc/heat/heat.conf  keystone_authtoken  admin_password
```

```
$Demo_User_Passwd
openstack-config --set /etc/heat/heat.conf ec2authtoken auth_uri http://
$Controller_Mgmt_IPAddress:5000/v2.0
openstack-config --set /etc/heat/heat.conf DEFAULT \
heat_metadata_server_url http://$Controller_Mgmt_IPAddress:8000
openstack-config --set /etc/heat/heat.conf DEFAULT \
heat_waitcondition_server_url http://$Controller_Mgmt_IPAddress:8000/v1/ waitcondition
#------------------------------ Ceilometer ------------------------------#
openstack-config --set /etc/ceilometer/ceilometer.conf publisher metering_
secret 000000
openstack-config --set /etc/ceilometer/ceilometer.conf DEFAULT \
rpc_backend ceilometer.openstack.common.rpc.impl_qpid
openstack-config --set /etc/ceilometer/ceilometer.conf DEFAULT auth_strategy
keystone
openstack-config --set /etc/ceilometer/ceilometer.conf keystone_authtoken \
auth_host $Controller_Mgmt_IPAddress
openstack-config --set /etc/ceilometer/ceilometer.conf keystone_authtoken admin_
user ceilometer
openstack-config --set /etc/ceilometer/ceilometer.conf keystone_authtoken admin_
tenant_name service
openstack-config --set /etc/ceilometer/ceilometer.conf keystone_authtoken auth_
protocol http
openstack-config --set /etc/ceilometer/ceilometer.conf keystone_authtoken \
auth_uri http://$Controller_Mgmt_IPAddress:5000
openstack-config --set /etc/ceilometer/ceilometer.conf keystone_authtoken \
admin_password $Demo_User_Passwd
openstack-config --set /etc/ceilometer/ceilometer.conf service_credentials \
os_auth_url http://$Controller_Mgmt_IPAddress:5000/v2.0
openstack-config --set /etc/ceilometer/ceilometer.conf service_credentials os_
username ceilometer
openstack-config --set /etc/ceilometer/ceilometer.conf service_credentials os_
tenant_name service
openstack-config --set /etc/ceilometer/ceilometer.conf service_credentials os_
password $Demo_User_Passwd
#--------------------------- Start Services ----------------------------#
service openstack-glance-api restart
service openstack-glance-registry restart
service openstack-nova-api restart
```

```
service openstack-nova-cert restart
service openstack-nova-consoleauth restart
service openstack-nova-scheduler restart
service openstack-nova-conductor restart
service openstack-nova-novncproxy restart
service neutron-server restart
service neutron-metadata-agent restart
service neutron-openvswitch-agent restart
service neutron-l3-agent restart
service neutron-dhcp-agent restart
service memcached restart
service httpd restart
service openstack-cinder-api restart
service openstack-cinder-scheduler restart
swift-init proxy restart
service rsyslog restart
service openstack-heat-api restart
service openstack-heat-api-cfn restart
service openstack-heat-engine restart
service openstack-ceilometer-api restart
service openstack-ceilometer-notification restart
service openstack-ceilometer-central restart
service openstack-ceilometer-collector restart
service openstack-ceilometer-alarm-evaluator restart
service openstack-ceilometer-alarm-notifier restart
#------------------------- Chkconfig services ---------------------------#
chkconfig openstack-glance-api on
chkconfig openstack-glance-registry on
chkconfig openstack-nova-api on
chkconfig openstack-nova-cert on
chkconfig openstack-nova-consoleauth on
chkconfig openstack-nova-scheduler on
chkconfig openstack-nova-conductor on
chkconfig openstack-nova-novncproxy on
chkconfig neutron-metadata-agent on
chkconfig neutron-server on
chkconfig neutron-openvswitch-agent on
chkconfig neutron-dhcp-agent on
```

```
chkconfig neutron-l3-agent on
chkconfig httpd on
chkconfig memcached on
chkconfig openstack-cinder-api on
chkconfig openstack-cinder-scheduler on
chkconfig openstack-swift-proxy on
chkconfig memcached on
chkconfig rsyslog on
chkconfig openstack-heat-api on
chkconfig openstack-heat-api-cfn on
chkconfig openstack-heat-engine on
chkconfig openstack-ceilometer-api on
chkconfig openstack-ceilometer-notification on
chkconfig openstack-ceilometer-central on
chkconfig openstack-ceilometer-collector on
chkconfig openstack-ceilometer-alarm-evaluator on
chkconfig openstack-ceilometer-alarm-notifier on
```

附录三 Xiandian_Install_Compute_Node.sh

计算节点安装脚本

```
#!/bin/bash
source ./xiandian_pre.sh
#--------------------------- Install Packages ----------------------------#
yum -y upgrade
service NetworkManager stop 2>/dev/null
chkconfig NetworkManager off 2>/dev/null
yum install -y openstack-utils openstack-selinux openssh-clients perl wget \
openssh-clients expect ntp openvswitch \
openstack-nova-compute openstack-utils openstack-selinux spice-html5 MySQL-
python \
openstack-neutron-ml2 openstack-neutron-openvswitch \
openstack-cinder scsi-target-utils openstack-utils \
openstack-swift openstack-swift-account openstack-swift-container \
openstack-swift-object xfsprogs xinetd rsync \
openstack-ceilometer-compute python-ceilometerclient python-pecan
sed -i 's/SELINUX=enforcing/SELINUX=permissive/g' /etc/selinux/config
setenforce 0  2>&1
yum upgrade -y
```

```
iptables -F
iptables -X
iptables -Z
service iptables save
sed  -i '/ntpdate/d' /etc/crontab
echo "*/1  *  *  * root  ntpdate controller" >>  /etc/crontab
netdate controller
chkconfig ntpdate on
/etc/init.d/crond restart
sed -i -e '/net.ipv4.ip_forward.*/d' -e '/net.ipv4.conf.all.rp_filter.*/d' \
-e '/net.ipv4.conf.default.rp_filter.*/d' /etc/sysctl.conf
cat >>/etc/sysctl.conf <<-EOF
net.ipv4.ip_forward=1
net.ipv4.conf.all.rp_filter=0
net.ipv4.conf.default.rp_filter=0
EOF
sysctl -p
mkfs.xfs -i size=1024 -f /dev/$Stroage_Swift_Disk
pvcreate /dev/$Stroage_Cinder_Disk
vgcreate cinder-volumes /dev/$Stroage_Cinder_Disk
cat <<EOF >> /etc/hosts
$Controller_Mgmt_IPAddress  $Controller_Hostname
$Compute_Mgmt_IPAddress     $Compute_Hostname
EOF
#------------------------- Configure the Network ------------------------#
service openvswitch start
chkconfig openvswitch on
sed -i 's/NM_CONTROLLED.*/NM_CONTROLLED=no/g' /etc/sysconfig/network-scripts/
ifcfg-*
sed  -i -e '/IPADDR/d' -e '/NETMASK/d' -e '/GATEWAY/d' -e '/DNS1/d' \
-e 's/ONBOOT=.*/ONBOOT=yes/g' -e 's/NM_CONTROLLED=.*/NM_CONTROLLED=no/g' \
/etc/sysconfig/network-scripts/ifcfg-eth*
cat > /etc/sysconfig/network-scripts/ifcfg-br-mgmt  << EOF
DEVICE=br-mgmt
IPADDR=$Compute_Mgmt_IPAddress
NETMASK=255.255.255.0
BOOTPROTO=static
GATEWAY=$Gateway_Mgmt
```

```
ONBOOT=yes
USERCTL=no
EOF
cat > /etc/sysconfig/network-scripts/ifcfg-br-ex  << EOF
DEVICE=br-ex
IPADDR=$Compute_External_IPAddress
NETMASK=255.255.255.0
BOOTPROTO=static
ONBOOT=yes
USERCTL=no
EOF
ovs-vsctl add-br br-mgmt
ovs-vsctl add-br br-eth0
ovs-vsctl add-port br-eth0 eth0
ovs-vsctl add-port br-mgmt "br-mgmt-br-eth0"
ovs-vsctl add-port br-eth0 "br-eth0-br-mgmt"
ovs-vsctl set interface  "br-eth0-br-mgmt" type=patch
ovs-vsctl set interface  "br-mgmt-br-eth0" type=patch
ovs-vsctl set interface "br-eth0-br-mgmt" options:peer="br-mgmt-br-eth0"
ovs-vsctl set interface "br-mgmt-br-eth0" options:peer="br-eth0-br-mgmt"
ovs-vsctl add-br br-eth1
ovs-vsctl add-br br-prv
ovs-vsctl add-br br-int
ovs-vsctl add-br br-ex
ovs-vsctl add-port br-eth1 eth1
ovs-vsctl add-port br-eth1 "br-eth1-br-int"
ovs-vsctl add-port br-eth1 "br-eth1-br-prv"
ovs-vsctl add-port br-eth1 "br-eth1-br-ex"
ovs-vsctl add-port br-prv "br-prv-br-eth1"
ovs-vsctl add-port br-int "br-int-br-eth1"
ovs-vsctl add-port br-ex "br-ex-br-eth1"
ovs-vsctl set interface  "br-prv-br-eth1" type=patch
ovs-vsctl set interface  "br-eth1-br-prv" type=patch
ovs-vsctl set interface  "br-eth1-br-ex" type=patch
ovs-vsctl set interface  "br-ex-br-eth1" type=patch
ovs-vsctl set interface  "br-eth1-br-int" type=patch
ovs-vsctl set interface  "br-int-br-eth1" type=patch
ovs-vsctl set interface "br-prv-br-eth1" options:peer="br-eth1-br-prv"
```

```
ovs-vsctl set interface "br-eth1-br-prv" options:peer="br-prv-br-eth1"
ovs-vsctl set interface "br-eth1-br-ex" options:peer="br-ex-br-eth1"
ovs-vsctl set interface "br-ex-br-eth1" options:peer="br-eth1-br-ex"
ovs-vsctl set interface "br-eth1-br-int" options:peer="br-int-br-eth1"
ovs-vsctl set interface "br-int-br-eth1" options:peer="br-eth1-br-int"
service network restart
#-------------------------- Configuration Nova -------------------------#
openstack-config --set /etc/nova/nova.conf DEFAULT auth_strategy keystone
openstack-config --set /etc/nova/nova.conf keystone_authtoken \
auth_uri http://$Controller_Mgmt_IPAddress:5000
openstack-config --set /etc/nova/nova.conf keystone_authtoken auth_host
$Controller_Mgmt_IPAddress
openstack-config --set /etc/nova/nova.conf keystone_authtoken auth_ protocol
http
openstack-config --set /etc/nova/nova.conf keystone_authtoken auth_port 35357
openstack-config --set /etc/nova/nova.conf keystone_authtoken admin_user nova
openstack-config --set /etc/nova/nova.conf keystone_authtoken admin_ tenant_
name service
openstack-config --set /etc/nova/nova.conf keystone_authtoken admin_ password
$Demo_User_Passwd
openstack-config --set /etc/nova/nova.conf DEFAULT rpc_backend qpid
openstack-config --set /etc/nova/nova.conf DEFAULT qpid_hostname $Controller_
Mgmt_IPAddress
openstack-config --set /etc/nova/nova.conf DEFAULT my_ip $Compute_Mgmt_
IPAddress
openstack-config --set /etc/nova/nova.conf DEFAULT vnc_enabled True
openstack-config --set /etc/nova/nova.conf DEFAULT vncserver_listen 0.0.0.0
openstack-config --set /etc/nova/nova.conf DEFAULT \
vncserver_proxyclient_address $Compute_Mgmt_IPAddress
openstack-config --set /etc/nova/nova.conf DEFAULT \
novncproxy_base_url http://$Controller_Mgmt_IPAddress:6080/vnc_auto.html
openstack-config --set /etc/nova/nova.conf DEFAULT glance_host $Controller_
Mgmt_IPAddress
sed -i "/INPUT.*REJECT/i\-A INPUT -m state --state NEW -m tcp -p tcp --dport
5900:5909 \
-j ACCEPT" /etc/sysconfig/iptables
openstack-config --set /etc/nova/nova.conf DEFAULT instance_usage_audit True
openstack-config --set /etc/nova/nova.conf DEFAULT instance_usage_audit_ period
```

```
hour
openstack-config --set /etc/nova/nova.conf DEFAULT notify_on_state_change
vm_and_task_state
openstack-config --set /etc/nova/nova.conf DEFAULT \
notification_driver nova.openstack.common.notifier.rpc_notifier
openstack-config --set /etc/nova/nova.conf DEFAULT nnotification_driver
ceilometer.compute.nova_notifier
sed -i 's/nnotification_driver/notification_driver/g' /etc/nova/nova.conf
openstack-config --set /etc/nova/nova.conf DEFAULT rpc_backend qpid
#------------------------- Configuration Neutron ----------------------#
openstack-config --set /etc/neutron/neutron.conf DEFAULT auth_strategy keystone
openstack-config --set /etc/neutron/neutron.conf DEFAULT \
rpc_backend neutron.openstack.common.rpc.impl_qpid
openstack-config --set /etc/neutron/neutron.conf DEFAULT qpid_hostname $Controller_
Mgmt_IPAddress
openstack-config --set /etc/neutron/neutron.conf DEFAULT core_plugin ml2
openstack-config --set /etc/neutron/neutron.conf DEFAULT service_plugins router
openstack-config --set /etc/neutron/neutron.conf DEFAULT control_exchange
neutron
openstack-config --set /etc/neutron/neutron.conf DEFAULT \
notification_driver neutron.openstack.common.notifier.rpc_notifier
openstack-config --set /etc/neutron/neutron.conf keystone_authtoken \
auth_uri http://$Controller_Mgmt_IPAddress:5000
openstack-config --set /etc/neutron/neutron.conf keystone_authtoken auth_ host
$Controller_Mgmt_IPAddress
openstack-config --set /etc/neutron/neutron.conf keystone_authtoken auth_
protocol http
openstack-config --set /etc/neutron/neutron.conf keystone_authtoken auth_ port
35357
openstack-config --set /etc/neutron/neutron.conf keystone_authtoken admin_
tenant_name service
openstack-config --set /etc/neutron/neutron.conf keystone_authtoken admin_ user
neutron
openstack-config --set /etc/neutron/neutron.conf keystone_authtoken admin_
password $Demo_User_Passwd
openstack-config --set /etc/neutron/neutron.conf DEFAULT notify_nova_ on_port_
status_changes True
openstack-config --set /etc/neutron/neutron.conf DEFAULT notify_nova_ on_port_
```

```
data_changes True
openstack-config --set /etc/neutron/plugins/ml2/ml2_conf.ini ml2 type_ drivers
vlan
openstack-config --set /etc/neutron/plugins/ml2/ml2_conf.ini ml2 tenant_ network_
types  vlan
openstack-config  --set  /etc/neutron/plugins/ml2/ml2_conf.ini  ml2  mechanism_
drivers  openvswitch
openstack-config --set /etc/neutron/plugins/ml2/ml2_conf.ini ml2_type_ vlan \
network_vlan_ranges physnet1:$Network_Start_Vlan_ID:$Network_End_Vlan_ID
openstack-config  --set  /etc/neutron/plugins/ml2/ml2_conf.ini  ovs  enable_
tunneling False
openstack-config  --set  /etc/neutron/plugins/ml2/ml2_conf.ini  ovs  integration_
bridge br-int
openstack-config  --set  /etc/neutron/plugins/ml2/ml2_conf.ini  ovs  bridge_
mappings physnet1:br-prv
openstack-config --set /etc/neutron/plugins/ml2/ml2_conf.ini securitygroup \
firewall_driver neutron.agent.linux.iptables_firewall.OVSHybridIptables FirewallDriver
openstack-config --set /etc/neutron/plugins/ml2/ml2_conf.ini securitygroup enable_
zsecurity_group True
ln -s plugins/ml2/ml2_conf.ini  /etc/neutron/plugin.ini
cp /etc/init.d/neutron-openvswitch-agent  /etc/init.d/neutron-openvswitch- agent.
orig
sed -i 's,plugins/openvswitch/ovs_neutron_plugin.ini,plugin.ini,g' /etc/init.d/
neutron- openvswitch-agent
openstack-config  --set  /etc/nova/nova.conf  DEFAULT  network_api_class  nova.
network.neutronv2.api.API
openstack-config  --set  /etc/nova/nova.conf  DEFAULT  neutron_url  http://
$Controller_Mgmt_IPAddress:9696
openstack-config  --set  /etc/nova/nova.conf  DEFAULT  neutron_auth_strategy
keystone
openstack-config --set /etc/nova/nova.conf DEFAULT neutron_admin_tenant_ name
service
openstack-config  --set  /etc/nova/nova.conf  DEFAULT  neutron_admin_username
neutron
openstack-config  --set  /etc/nova/nova.conf  DEFAULT  neutron_admin_password
$Demo_User_Passwd
openstack-config --set /etc/nova/nova.conf DEFAULT \
neutron_admin_auth_url http://$Controller_Mgmt_IPAddress:35357/v2.0
```

```
openstack-config --set /etc/nova/nova.conf DEFAULT \
linuxnet_interface_driver nova.network.linux_net.LinuxOVSInterfaceDriver
openstack-config --set /etc/nova/nova.conf DEFAULT firewall_driver nova.
virt.firewall.NoopFirewallDriver
openstack-config --set /etc/nova/nova.conf DEFAULT security_group_api neutron
openstack-config --set /etc/nova/nova.conf DEFAULT vif_plugging_is_fatal False
openstack-config --set /etc/nova/nova.conf DEFAULT vif_plugging_timeout 10
openstack-config --set /etc/nova/nova.conf DEFAULT service_neutron_metadata_
proxy true
openstack-config --set /etc/nova/nova.conf DEFAULT neutron_metadata_ proxy_
shared_secret 000000
openstack-config --set /etc/nova/nova.conf DEFAULT instance_usage_audit True
openstack-config --set /etc/nova/nova.conf DEFAULT instance_usage_audit_ period
hour
openstack-config --set /etc/nova/nova.conf DEFAULT notify_on_state_change vm_
and_task_state
sed -i 's/notification_driver.*//g' /etc/nova/nova.conf
openstack-config --set /etc/nova/nova.conf DEFAULT \
notification_driver nova.openstack.common.notifier.rpc_notifier
sed -i 's/notification_driver.*/notification_driver = \
nova.openstack.common.notifier.rpc_notifier\nnotification_driver = ceilometer.
compute.nova_notifier/g' \
/etc/nova/nova.conf
#------------------------- Configuration Cinder -------------------------#
openstack-config --set /etc/cinder/cinder.conf DEFAULT auth_strategy keystone
openstack-config --set /etc/cinder/cinder.conf keystone_authtoken \
auth_uri http://$Controller_Mgmt_IPAddress:5000
openstack-config --set /etc/cinder/cinder.conf keystone_authtoken auth_host
$Controller_Mgmt_IPAddress
openstack-config --set /etc/cinder/cinder.conf keystone_authtoken auth_protocol http
openstack-config --set /etc/cinder/cinder.conf keystone_authtoken auth_port 35357
openstack-config --set /etc/cinder/cinder.conf keystone_authtoken admin_user
cinder
openstack-config --set /etc/cinder/cinder.conf keystone_authtoken admin_
tenant_name service
openstack-config --set /etc/cinder/cinder.conf keystone_authtoken admin_
password $Demo_User_Passwd
openstack-config --set /etc/cinder/cinder.conf DEFAULT \
```

```
rpc_backend cinder.openstack.common.rpc.impl_qpid
openstack-config --set /etc/cinder/cinder.conf DEFAULT qpid_hostname $Controller_
Mgmt_IPAddress
openstack-config --set /etc/cinder/cinder.conf DEFAULT control_exchange cinder
openstack-config --set /etc/cinder/cinder.conf DEFAULT \
notification_driver cinder.openstack.common.notifier.rpc_notifier
openstack-config --set /etc/cinder/cinder.conf database \
connection mysql://cinder:$Demo_DB_Passwd@$Controller_Mgmt_IPAddress/cinder
openstack-config --set /etc/cinder/cinder.conf DEFAULT glance_host $Controller_
Mgmt_IPAddress
openstack-config --set /etc/cinder/cinder.conf DEFAULT iscsi_helper tgtadm
echo "include /etc/cinder/volumes/*" >>/etc/tgt/targets.conf
#------------------------ Configuration Swift -------------------------#
echo "/dev/${!OBJECT_DISK} /swift/node xfs loop,noatime,nodiratime,nobarrier,
logbufs=8 0 0" >> /etc/fstab
mkdir -p /swift/node
mount /swift/node
chown -R swift:swift /swift
mkdir -p /etc/swift
scp $Controller_Hostname:/etc/swift/swift.conf /etc/swift/
cat <<EOF > /etc/rsync.conf
uid = swift
gid = swift
log file = /var/log/rsyncd.log
pid file = /var/run/rsyncd.pid
address = 127.0.0.1
[account]
max connections = 2
path = /swift/node
read only = false
lock file = /var/lock/account.lock
[container]
max connections = 2
path = /swift/node
read only = false
lock file = /var/lock/container.lock
[object]
max connections = 2
```

```
path = /swift/node
read only = false
lock file = /var/lock/object.lock
EOF
sed -i 's/yes/no/' /etc/xinetd.d/rsync
sed -i 's/IPv6/IPv4/' /etc/xinetd.d/rsync
service xinetd restart
chkconfig xinetd on
sed -i 's/^bind_ip.*/#bind_ip = /g' /etc/swift/account-server.conf
sed -i 's/^bind_ip.*/#bind_ip = /g' /etc/swift/container-server.conf
sed -i 's/^bind_ip.*/#bind_ip = /g' /etc/swift/object-server.conf
mv /etc/swift/account-server.conf /etc/swift/account-server/1.conf
mv /etc/swift/container-server.conf /etc/swift/container-server/1.conf
mv /etc/swift/object-server.conf /etc/swift/object-server/1.conf
openstack-config --set /etc/swift/account-server/1.conf DEFAULT bind_port 6002
openstack-config --set /etc/swift/account-server/1.conf DEFAULT workers 2
openstack-config --set /etc/swift/account-server/1.conf DEFAULT user swift
openstack-config --set /etc/swift/account-server/1.conf DEFAULT devices
/swift/node
openstack-config --set /etc/swift/account-server/1.conf DEFAULT mount_ check
false
openstack-config --set /etc/swift/account-server/1.conf DEFAULT log_name
swift-account
openstack-config --set /etc/swift/account-server/1.conf DEFAULT log_facility
LOG_LOCAL1
openstack-config --set /etc/swift/account-server/1.conf DEFAULT log_level DEBUG
openstack-config --set /etc/swift/account-server/1.conf account-replicator vm_
test_mode no
openstack-config --set /etc/swift/container-server/1.conf DEFAULT bind_ port
6001
openstack-config --set /etc/swift/container-server/1.conf DEFAULT workers 2
openstack-config --set /etc/swift/container-server/1.conf DEFAULT user swift
openstack-config --set /etc/swift/container-server/1.conf DEFAULT devices
/swift/node
openstack-config --set /etc/swift/container-server/1.conf DEFAULT mount_ check
false
openstack-config --set /etc/swift/container-server/1.conf DEFAULT log_ name
swift-container
```

```
openstack-config --set /etc/swift/container-server/1.conf DEFAULT log_ facility
LOG_LOCAL1
openstack-config --set /etc/swift/container-server/1.conf DEFAULT log_ level
DEBUG
openstack-config --set /etc/swift/container-server/1.conf container- replicator
vm_test_mode no
openstack-config --set /etc/swift/object-server/1.conf DEFAULT bind_port 6000
openstack-config --set /etc/swift/object-server/1.conf DEFAULT workers 2
openstack-config --set /etc/swift/object-server/1.conf DEFAULT user swift
openstack-config --set /etc/swift/object-server/1.conf DEFAULT devices /swift/
node
openstack-config --set /etc/swift/object-server/1.conf DEFAULT mount_ check
false
openstack-config --set /etc/swift/object-server/1.conf DEFAULT log_name swift-
object
openstack-config --set /etc/swift/object-server/1.conf object-replicator vm_
test_mode no
echo "local1.* /var/log/swift/storage1.log">>/etc/rsyslog.d/10-swift.conf
mkdir -p /var/log/swift
chown -R swift:swift /var/log/swift
service rsyslog start
chkconfig rsyslog on
scp $Controller_Hostname:/etc/swift/*.ring.gz /etc/swift/
mkdir -p /var/swift/recon
chown -R swift:swift /var/swift
chown -R swift:swift /etc/swift
chown -R swift:swift /var/cache/swift
#---------------------- Configuration Ceilometer ----------------------#
openstack-config --set /etc/ceilometer/ceilometer.conf DEFAULT auth_ strategy
keystone
openstack-config --set /etc/ceilometer/ceilometer.conf publisher metering_
secret 000000
openstack-config --set /etc/ceilometer/ceilometer.conf DEFAULT \
rpc_backend ceilometer.openstack.common.rpc.impl_qpid
openstack-config --set /etc/ceilometer/ceilometer.conf DEFAULT \
qpid_hostname $Controller_Mgmt_IPAddress
openstack-config --set /etc/ceilometer/ceilometer.conf keystone_authtoken \
auth_host $Controller_Mgmt_IPAddress
```

```
openstack-config --set /etc/ceilometer/ceilometer.conf keystone_authtoken admin_
user ceilometer
openstack-config --set /etc/ceilometer/ceilometer.conf keystone_authtoken admin_
tenant_name service
openstack-config --set /etc/ceilometer/ceilometer.conf keystone_authtoken auth_
protocol http
openstack-config --set /etc/ceilometer/ceilometer.conf keystone_authtoken \
auth_uri http://$Controller_Mgmt_IPAddress:5000
openstack-config --set /etc/ceilometer/ceilometer.conf keystone_authtoken \
admin_password $Demo_User_Passwd
openstack-config --set /etc/ceilometer/ceilometer.conf service_credentials os_
username ceilometer
openstack-config --set /etc/ceilometer/ceilometer.conf service_credentials os_
tenant_name service
openstack-config --set /etc/ceilometer/ceilometer.conf service_credentials os_
password $Demo_User_Passwd
openstack-config --set /etc/ceilometer/ceilometer.conf service_credentials \
os_auth_url http://$Controller_Mgmt_IPAddress:5000/v2.0
#------------------------------ Start Services --------------------------#
for service in openstack-swift-object openstack-swift-object-replicator
openstack-swift-object-updater \
openstack-swift-object-auditor openstack-swift-container openstack-swift- container-
replicator \
openstack-swift-container-updater openstack-swift-container-auditor openstack-
swift-account \
openstack-swift-account-replicator openstack-swift-account-reaper openstack- swift-
account-auditor; \
do chkconfig $service on; done
chkconfig openstack-cinder-volume on
chkconfig tgtd on
chkconfig openstack-ceilometer-compute on
chkconfig libvirtd on
chkconfig messagebus on
chkconfig openstack-nova-compute on
service openstack-cinder-volume restart
service tgtd restart
service openstack-ceilometer-compute restart
service iptables reload
```

```
service openstack-nova-compute restart
service libvirtd restart
service messagebus restart
service openstack-nova-compute restart
service neutron-metadata-agent restart
service neutron-openvswitch-agent restart
swift-init object-server restart
swift-init object-replicator restart
swift-init object-updater restart
swift-init object-auditor restart
swift-init container-server restart
swift-init container-replicator restart
swift-init container-updater restart
swift-init container-auditor restart
swift-init account-server restart
swift-init account-replicator restart
swift-init account-auditor restart
```

附录四 Keystone-manage-tenant.sh

Keystone 租户管理

```
#!/bin/bash
if [  -f "/etc/keystone/admin-openrc.sh" ];then
        source /etc/keystone/admin-openrc.sh
else
env_path='find / -name admin-openrc.sh'
        source $env_path
fi
    echo -e "\033[31mPlease Input new tenant name : eg (openstack)\033[0m "
        read New_Tenant_Name
if [ ! -n "$New_Tenant_Name" ];then
            echo -e "\033[31mTenant Name Is Empty,Exit\033[0m "
            exit 1
        fi
    echo  -e  "\033[31mPlease  Input  tenant  description  :  eg  (openstack 
description)\033[0m "
        read New_Tenant_des
if [ ! -n "$New_Tenant_des" ];then
            echo -e "\033[31mTenant  Description  Is Empty,Exit\033[0m "
```

```
        exit 1
fi
T_Start='echo $New_Tenant_Range |awk -F- '{ print $1}'| awk '{print $0+0}''
N_Start='printf "%03d\n" $T_Start'
T_End='echo $New_Tenant_Range |awk -F- '{ print $2}' | awk '{print $0+0}''
N_End='printf "%03d\n" $T_End'
      T_End1=$[$T_End+1]
            keystone  tenant-create  --name  $New_Tenant_Name  --description
$New_Tenant_des
            echo -e "\033[31mKeystone All Tenant List\033[0m "
            keystone tenant-list
```

附录五 Keystone-manage-user.sh

Keystone 用户管理

```
#!/bin/bash
if [  -f "/etc/keystone/admin-openrc.sh" ];then
      source /etc/keystone/admin-openrc.sh
else
env_path='find / -name admin-openrc.sh'
      source $env_path
fi
      echo -e "\033[31mPlease Input New User Name : eg (username)\033[0m "
      read New_User_Name
if [ ! -n "$New_User_Name" ];then
              echo -e "\033[31mUser Name Is Empty,Exit\033[0m "
              exit 1
 fi
      echo -e "\033[31mPlease Input User Password: eg (000000)\033[0m "
      read New_User_Pw
if [ ! -n "$New_User_Pw" ];then
            echo -e "\033[31mPasswd Is Empty,Exit\033[0m "
            exit 1
         fi
      echo -e "\033[31mPlease Input User Email Address,\
If don't need  press enter: eg (openstack.com)\033[0m "
      read New_User_Email
if [ ! -n "$New_User_Email" ];then
```

```
            echo -e "\033[31mEmail Address Is Empty,Exit\033[0m "
            exit 1
        fi
   echo  -e  "\033[31mPlease  Input  User    Beginning  And  End   Number:  eg
(001-002)\033[0m "
       read New_User_Range
if [ ! -n "$New_User_Range" ];then
                echo -e "\033[31mNumber Is Empty,Exit\033[0m "
                exit 1
            else
U_Start='echo $New_User_Range |awk -F- '{ print $1}'| awk '{print $0+0}''
N_U_Start='printf "%03d\n" $U_Start'
U_End='echo $New_User_Range |awk -F- '{ print $2}' | awk '{print $0+0}''
N_U_End='printf "%03d\n" $U_End'
                U_End1=$[$U_End+1]
IF_username_exists='keystone user-list | sed '1,3d'|sed '$d'|awk '{print $4}''
                    for username_exists in $IF_username_exists;do
                        for (( username_number = \
$U_Start;username_number< $U_End1;username_number++ ));do
real_username_number='printf "%03d\n" $username_number'
                            if [ $New_User_Name$real_username_number == \
$username_exists ];then
                                echo -e "\033[31mUser \
$New_User_Name$real_username_number is exists\033[0m "
                                exit 1
                            fi
done
                    done
        fi
       echo -e "\033[31mPlease enter the User belong Roles Name, \
Press enter for '_member_' role by default: eg (admin)\033[0m "
       read New_User_Role
if [ ! -n "$New_User_Role" ];then
New_User_Role=_member_
        else
IF_Role_Exists='keystone role-list |sed '1,3d' |sed '$d' |awk '{print $4}''
                if  echo "${IF_Role_Exists[@]}" | grep -w "$New_User_Role" >>
/dev/null ; then
```

```
                    echo "exists" >> /dev/null
                else
                        echo -e "\033[31mRole $New_User_Role not exists\033[0m "
                        exit 1
                fi
        fi

    echo -e "\033[31mPlease Input User belong Tenant Name: eg (tenantname)\033[0m "
        read New_User_Tenant
if [ ! -n "$New_User_Tenant" ];then
                        echo -e "\033[31mTenant Name Is Empty,Exit\033[0m "
                        exit 1
        else
IF_Tenant_Exists='keystone tenant-list |sed '1,3d' |sed '$d' |awk '{print $4}''
                if  echo "${IF_Tenant_Exists[@]}" | grep -w "$New_User_ Tenant"
>> /dev/null ; then
                    echo "exists" >> /dev/null
                else
                        echo  -e  "\033[31mTenant  $New_User_Tenant  not  exists\
033[0m "
                        exit 1
                fi
        fi
                        for (( username_number = $U_Start;username_number<\
$U_End1;username_number++ ));do
real_username_number='printf "%03d\n" $username_number'
                                keystone user-create \
--name $New_User_Name$real_username_number --pass $New_User_Pw \
--email $New_User_Name$real_username_number@$New_User_Email
                                keystone user-role-add \
--user  $New_User_Name$real_username_number  --role  $New_User_Role  --tenant
$New_User_Tenant
                        done
                                echo -e "\033[31mKeystone All User List\033[0m "
                                keystone user-list
```

附录六　Keystone-manage-add-role.sh

Keystone 权限管理

```
#!/bin/bash
#-------------------------------- 1st keystone -------------------------------
if [  -f "/etc/keystone/admin-openrc.sh" ];then
        source /etc/keystone/admin-openrc.sh
else
env_path='find / -name admin-openrc.sh'
        source $env_path
fi
        echo -e "\033[31mPlease Enter The User Name\033[0m "
        read Add_Role_Username
        echo  -e  "\033[31mPlease  Input  User   Beginning  And  End   Number:  eg
(001-002)\033[0m "
          read Add_User_Range
if [ ! -n "$Add_User_Range" ];then
Add_User_Range=$Add_User_Range
                  else
A_R_Start='echo $Add_User_Range |awk -F- '{ print $1}'| awk '{print $0+0}''
A_R_U_Start='printf "%03d\n" $A_R_Start'
A_R_End='echo $Add_User_Range |awk -F- '{ print $2}' | awk '{print $0+0}''
A_R_U_End='printf "%03d\n" $A_R_End'
                      A_R_End1=$[$A_R_End+1]
              fi
        echo -e "\033[31mPlease Enter the Tenant Name\033[0m "
                read Add_Role_Tenant
IF_Tenant_Exists='keystone tenant-list |sed '1,3d' |sed '$d' |awk '{print $4}''
                if  echo "${IF_Tenant_Exists[@]}" | grep -w "$Add_Role_ Tenant" >>
/dev/null ; then
                        echo "exists" >> /dev/null
                else
                        echo -e "\033[31mTenant $Add_Role_Tenant not exists\ 033[0m "
                        exit 1
                fi
         echo -e "\033[31mPlease Enter the  Role Name\033[0m "
                 read Add_Role_New_Role
IF_Role_Exists='keystone role-list |sed '1,3d' |sed '$d' |awk '{print $4}''
```

```
            if  echo "${IF_Role_Exists[@]}" | grep -w "$Add_Role_New_ Role"
>> /dev/null ; then
                    echo "exists" >> /dev/null
            else
                    echo -e "\033[31mRole $Add_Role_New_Role not exists\033 [0m "
                    exit 1
            fi
            for
(( username_number=$A_R_Start;username_number<$A_R_End1; username_number++ ));do
real_username_number='printf "%03d\n" $username_number'
                    keystone user-role-add --user $Add_Role_Username$ real_
username_number \
--role $Add_Role_New_Role --tenant  $Add_Role_Tenant
                    echo -e "\033[31mKeystone user \
$Add_Role_Username$real_username_number tenant $Add_Role_Tenant role list\
033[0m "
                    keystone user-role-list --user $Add_Role_Username$ real_
username_number \
--tenant  $Add_Role_Tenant
            done
```

附录七　qpid-tool.txt

QPID 消息情况

```
[root@controller ~]# qpid-tool 172.24.2.10
Management Tool for QPID
qpid: list exchange
Object Summary:
    ID   Created   Destroyed  Index
    ===================================================================
    442  04:13:39  -          870.
    443  06:23:40  -          870.None
    444  04:13:39  -          870.amq.direct
    445  04:13:39  -          870.amq.fanout
    446  04:13:39  -          870.amq.match
    447  04:13:39  -          870.amq.topic
    448  06:53:10  -          870.ceilometer
```

```
449  06:53:10  -         870.ceilometer.agent.notification_fanout
450  06:53:10  -         870.ceilometer.alarm_fanout
451  06:53:10  -         870.ceilometer.collector_fanout
452  04:18:12  -         870.cert_fanout
453  06:53:10  -         870.cinder
454  06:16:53  -         870.cinder-backup_fanout
455  07:36:54  -         870.cinder-scheduler_fanout
456  06:39:45  -         870.cinder-volume_fanout
457  04:20:38  -         870.compute_fanout
458  04:18:12  -         870.conductor_fanout
459  04:18:12  -         870.consoleauth_fanout
460  04:22:37  -         870.dhcp_agent_fanout
461  06:53:10  -         870.glance
462  06:53:10  -         870.heat
463  04:22:37  -         870.l3_agent_fanout
464  04:22:37  -         870.neutron
465  04:18:12  -         870.nova
466  06:20:49  -         870.openstack
467  04:31:58  -         870.q-agent-notifier-network-delete_fanout
468  04:31:58  -         870.q-agent-notifier-port-update_fanout
469  04:31:58  -         870.q-agent-notifier-security_group-update_fanout
470  04:31:58  -         870.q-agent-notifier-tunnel-update_fanout
471  04:13:39  -         870.qmf.default.direct
472  04:13:39  -         870.qmf.default.topic
473  04:13:39  -         870.qpid.management
474  06:54:58  -         870.reply_0095c2a1d0e04080b229ffb765f71ebb
475  04:35:47  -         870.reply_0117628681ac4dab99530934a7b26cc6
476  04:22:37  -         870.reply_0117ac258baa4bfeb685bcf9133fea09
477  06:06:38  -         870.reply_0142c83ee1964b73a633ab387fe11bee
478  06:01:56  -         870.reply_130fc2b6689e4e32b8b8768d67fa15ed
479  06:26:09  -         870.reply_1b98d8a2015341dcbcbe7d8e3abeced0
480  04:17:12  -         870.reply_21e269f5f0e64c438217fde55a831366
481  06:01:44  -         870.reply_234310add5de446e8555c9e47fa59d24
482  03:19:23  -         870.reply_24339be36c014f1a8ed9604b207bea31
483  06:55:03  -         870.reply_25c9a95bd5564096b8cb58bcd9af33cb
484  06:25:27  -         870.reply_281d017d69a94d6fb4e8e6344a60985e
485  06:08:11  -         870.reply_28deb16d7b1042799d49cac822ad69c6
```

```
486  06:23:43  -        870.reply_2b581ecba2164a00bcbab29403866f69
487  04:17:12  -        870.reply_2b8d0e498a9543f288b619f8fa5d0f0b
488  06:35:28  -        870.reply_3027735d3ad7498cbfe04df793b26b8f
489  06:01:44  -        870.reply_331ffe6edb1d47bda45f0ea4a81d3628
490  02:45:28  -        870.reply_3e4ece4c2b2945dcae5c1e7cf0545f8f
491  04:22:57  -        870.reply_4144f545225b4dbfae3370c788da3c68
492  04:17:24  -        870.reply_4d98067fa9b94ab881eeabf9de9f94e4
493  06:53:35  -        870.reply_507d94e4f8d14cb4beaf43bf5bfbfd61
494  06:36:01  -        870.reply_582eeb26654249c3b59c1cfb1d0846a8
495  04:31:58  -        870.reply_63c9c83547e34f17ae45eb7a029e8477
496  06:54:58  -        870.reply_64b8deb58c0a406ca1f2a6271bd625f0
497  04:31:58  -        870.reply_65278e7e02dd4c04accdd032e4fa2884
498  06:08:10  -        870.reply_666196afdab8475bade0cd040589b770
499  06:24:14  -        870.reply_6cfb8f51d5944621be775d93f1029951
500  03:19:31  -        870.reply_727a75d7bd2e4aa8b83fb741126159a9
501  04:22:57  -        870.reply_75e7365155ed4694bb7450c68d632475
502  04:23:21  -        870.reply_7e8b0a1b65824e95942be1fb76dac5a5
503  06:06:38  -        870.reply_85e9741a3bc3413f95bd5fe7675f7182
504  06:01:56  -        870.reply_8c1dc2ef5a6645aabf286ae5bab9d7da
505  06:36:01  -        870.reply_911baf796d374a529ba97a4855b1a8e6
506  02:45:29  -        870.reply_95f81503fec946ce8bb72c55b5884036
507  04:17:51  -        870.reply_9e0a28982da54608943305b9b2e4359a
508  04:22:37  -        870.reply_a0177825dc154bfabfb73a98df0daf74
509  03:18:34  -        870.reply_a9312ad8a9524d4ca1a5a85d0101b17a
510  06:26:09  -        870.reply_a9abf8a99fe54da1986d16928f344d47
511  06:25:28  -        870.reply_ab2ff012d8434c539b5b3983341b4ee1
512  04:17:50  -        870.reply_ac14c30c9eb3455284c17ad1b0a5b5c9
513  04:20:36  -        870.reply_b07694a7882b439f942f79ba6829aac3
514  06:25:26  -        870.reply_b5c8da117a0d4068977892a9bdcd2a76
515  06:55:03  -        870.reply_b60c834898a24e16a415113047b97118
516  03:18:34  -        870.reply_ba419dfad2c543d3865028beac3c607a
517  04:17:11  -        870.reply_be4b733bb4824e77847715991a602e36
518  04:22:37  -        870.reply_bf5e0d79340f4bf0be69be565d772c1d
519  06:01:56  -        870.reply_bfaaed56ae584a99ac3b605eddd85f70
520  06:06:09  -        870.reply_cbec16cd1fc04f34865446dbc6c2f375
521  06:25:28  -        870.reply_d19f578f7fc64f7fac96a79300644d1b
522  06:35:28  -        870.reply_e17d5ef1388c4c56b23ebd148ba60169
523  04:15:39  -        870.reply_e71d2ae08d534270833ae101c73b6cf0
```

```
    524  06:32:40  -          870.reply_e7e7b8f2632b4db797ec327d7d5cb9cd
    525  04:35:47  -          870.reply_f42ce978c1ad422ea2af37a90704ba54
    526  04:31:58  -          870.reply_f49c48ed0e6e4a34b31a1387f48917e2
    527  06:01:44  -          870.reply_f7e4092b4b8e416990b5ad0366f0a5e6
    528  04:18:12  -          870.scheduler_fanout
qpid: quit
Exiting...
```

附录八　nova –debug.txt

Nova 的故障排查反馈信息

```
REQ:  curl-i'http:  //172.24.0.10:  35357/v2.0/tokens'-XPOST-H"Content-Type:
application/json"-H"Accept: application/json"-H"User-Agent: python-novaCLIent"-d'{
    "auth": {
        "tenantName": "admin",
        "passwordCredentials": {
            "username": "admin",
            "password": "000000"
        }
    }
}'
REQ: curl-i'http: //172.24.2.10: 8774/v2/219c95eac5694e45bd5c7304613835 d3/
images/cfaf8992-bbb9-4895-bda1-a1547f0fa356'-XGET-H"X-
Auth-Project-Id: admin"-H"User-Agent: python-novaCLIent"-H"Accept: application/
json"-H"X-Auth-Token:  $token"REQ:  curl-i'http:  //172.24.2.10:  8774/v2/
219c95eac5694e45bd5c7304613835d3/flavors/3'-XGET-H"X-Auth-Project-Id: admin"
-H"User-
Agent: python-novaCLIent"-H"Accept: application/json"-H"X-Auth-Token: $token
"REQ: curl-i'http: //172.24.2.10: 8774/v2/219c95eac5694e45bd5c7304613835d3/ servers'
-XPOST-H"X-Auth-Project-Id: admin"-H"User-
Agent: python-novaCLIent"-H"Content-Type: application/json"-H"Accept: application/
json"-H"X-Auth-Token: $token"-d'{
    "server": {
        "name": "test1",
        "imageRef": "cfaf8992-bbb9-4895-bda1-a1547f0fa356",
        "flavorRef": "3",
        "max_count": 1,
        "min_count": 1,
        "networks": [
```

```
            {
                "uuid": "54cf9939-4d75-4ba2-9b27-6becf8f13561"
            }
        ]
    }
}'
REQ: curl-i'http: //172.24.2.10: 8774/v2/219c95eac5694e45bd5c7304613835d3/ servers/
a5843b51-ed0e-4271-a9c0-38757ebed798'-XGET-H"X-
Auth-Project-Id: admin"-H"User-Agent: python-novaCLIent"-H"Accept: application/ json"-
H"X-Auth-Token:  $token"\REQ:  curl-i'http:  //172.24.2.10:  8774/v2/  219c95
eac5694e45bd5c7304613835d3/flavors/3'-XGET-H"X-Auth-Project-Id:
admin"-H"User-
Agent: python-novaCLIent"-H"Accept: application/json"-H"X-Auth-Token: $token
"REQ:  curl-i'http:  //172.24.2.10:  8774/v2/219c95eac5694e45bd5c7304613835d3/
images/cfaf8992-bbb9-4895-bda1-a1547f0fa356'-XGET-H"X-
Auth-Project-Id: admin"-H"User-Agent: python-novaCLIent"-H"Accept: application/
json"-H"X-Auth-Token: $token"
```

附录九　virsh–list.txt

虚拟机信息

```
[root@compute ~]# virsh list --all
 Id    Name                           State
----------------------------------------------------
 3     instance-00000003              running
 4     instance-00000004              running

[root@compute ~]# virsh edit instance-00000003
<domain type='kvm'>
<name>instance-00000003</name>
<uuid>b46418fb-106f-4438-8b5a-567fd967c861</uuid>
<memory unit='KiB'>1048576</memory>
<currentMemory unit='KiB'>1048576</currentMemory>
<vcpu placement='static'>1</vcpu>
<sysinfo type='smbios'>
<system>
<entry name='manufacturer'>RDO Project</entry>
<entry name='product'>openstack Nova</entry>
<entry name='version'>2014.1.3-3.el6</entry>
```

```
<entry name='serial'>00000000-0000-0000-0000-0cc47a4db2a0</entry>
<entry name='uuid'>b46418fb-106f-4438-8b5a-567fd967c861</entry>
</system>
</sysinfo>
<os>
<type arch='x86_64' machine='rhel6.5.0'>hvm</type>
<boot dev='hd'/>
<smbios mode='sysinfo'/>
</os>
<features>
<acpi/>
<apic/>
</features>
<cpu mode='host-model'>
<model fallback='allow'/>
</cpu>
<clock offset='utc'>
<timer name='pit' tickpolicy='delay'/>
<timer name='rtc' tickpolicy='catchup'/>
<timer name='hpet' present='no'/>
</clock>
<on_poweroff>destroy</on_poweroff>
<on_reboot>restart</on_reboot>
<on_crash>destroy</on_crash>
<devices>
<emulator>/usr/libexec/qemu-kvm</emulator>
<disk type='file' device='disk'>
<driver name='qemu' type='qcow2' cache='none'/>
<source file='/var/lib/nova/instances/b46418fb-106f-4438-8b5a-567fd967c861/ disk'/>
<target dev='vda' bus='virtio'/>
<address type='pci' domain='0x0000' bus='0x00' slot='0x04' function='0x0'/>
</disk>
<controller type='usb' index='0'>
<address type='pci' domain='0x0000' bus='0x00' slot='0x01' function= '0x2'/>
</controller>
<interface type='bridge'>
<mac address='fa:16:3e:51:48:5f'/>
```

```
<source bridge='qbr33f37054-b6'/>
<target dev='tap33f37054-b6'/>
<model type='virtio'/>
<address type='pci' domain='0x0000' bus='0x00' slot='0x03' function= '0x0'/>
</interface>
<serial type='file'>
<source path='/var/lib/nova/instances/b46418fb-106f-4438-8b5a-567fd967c861/ console.
log'/>
<target port='0'/>
</serial>
<serial type='pty'>
<target port='1'/>
</serial>
<console type='file'>
<source path='/var/lib/nova/instances/b46418fb-106f-4438-8b5a-567fd967c861/ console.
log'/>
<target type='serial' port='0'/>
</console>
<input type='tablet' bus='usb'/>
<input type='mouse' bus='ps2'/>
<graphics type='vnc' port='-1' autoport='yes' listen='0.0.0.0' keymap= 'en-us'>
<listen type='address' address='0.0.0.0'/>
</graphics>
<video>
<model type='cirrus' vram='9216' heads='1'/>
<address type='pci' domain='0x0000' bus='0x00' slot='0x02' function= '0x0'/>
</video>
<memballoon model='virtio'>
<address type='pci' domain='0x0000' bus='0x00' slot='0x05' function= '0x0'/>
</memballoon>
</devices>
</domain>
```

附录十 vm_conf.txt

虚拟机配置信息

```
<domain type='kvm'>
<name>instance-00000003</name>
<uuid>b46418fb-106f-4438-8b5a-567fd967c861</uuid>
```

```
<memory unit='KiB'>1048576</memory>
<currentMemory unit='KiB'>1048576</currentMemory>
<vcpu placement='static'>1</vcpu>
<sysinfo type='smbios'>
<system>
<entry name='manufacturer'>RDO Project</entry>
<entry name='product'>OpenStack Nova</entry>
<entry name='version'>2014.1.3-3.el6</entry>
<entry name='serial'>00000000-0000-0000-0000-0cc47a4db2a0</entry>
<entry name='uuid'>b46418fb-106f-4438-8b5a-567fd967c861</entry>
</system>
</sysinfo>
<os>
<type arch='x86_64' machine='rhel6.5.0'>hvm</type>
<boot dev='hd'/>
<smbios mode='sysinfo'/>
</os>
<features>
<acpi/>
<apic/>
</features>
<cpu mode='host-model'>
<model fallback='allow'/>
</cpu>
<clock offset='utc'>
<timer name='pit' tickpolicy='delay'/>
<timer name='rtc' tickpolicy='catchup'/>
<timer name='hpet' present='no'/>
</clock>
<on_poweroff>destroy</on_poweroff>
<on_reboot>restart</on_reboot>
<on_crash>destroy</on_crash>
<devices>
<emulator>/usr/libexec/qemu-kvm</emulator>
<disk type='file' device='disk'>
<driver name='qemu' type='qcow2' cache='none'/>
<source file='/var/lib/nova/instances/b46418fb-106f-4438-8b5a-567fd967c861/ disk'/>
```

```
<target dev='vda' bus='virtio'/>
<address type='pci' domain='0x0000' bus='0x00' slot='0x04' function='0x0'/>
</disk>
<controller type='usb' index='0'>
<address type='pci' domain='0x0000' bus='0x00' slot='0x01' function='0x2'/>
</controller>
<interface type='bridge'>
<mac address='fa:16:3e:51:48:5f'/>
<source bridge='qbr33f37054-b6'/>
<target dev='tap33f37054-b6'/>
<model type='virtio'/>
<address type='pci' domain='0x0000' bus='0x00' slot='0x03' function='0x0'/>
</interface>
<serial type='file'>
<source path='/var/lib/nova/instances/b46418fb-106f-4438-8b5a-567fd967c861/ console.
log'/>
<target port='0'/>
</serial>
<serial type='pty'>
<target port='1'/>
</serial>
<console type='file'>
<source path='/var/lib/nova/instances/b46418fb-106f-4438-8b5a-567fd967c861/ console.
log'/>
<target type='serial' port='0'/>
</console>
<input type='tablet' bus='usb'/>
<input type='mouse' bus='ps2'/>
<graphics type='vnc' port='-1' autoport='yes' listen='0.0.0.0' keymap= 'en-us'>
<listen type='address' address='0.0.0.0'/>
</graphics>
<video>
<model type='cirrus' vram='9216' heads='1'/>
<address type='pci' domain='0x0000' bus='0x00' slot='0x02' function= '0x0'/>
</video>
<memballoon model='virtio'>
<address type='pci' domain='0x0000' bus='0x00' slot='0x05' function= '0x0'/>
```

```
</memballoon>
</devices>
</domain>
```

附录十一　mysql_full_bk.sh

数据库全备份脚本

```
#!/bin/bash
today='date +"%Y_%m_%d"'
bkdir=/opt/mysql/backup/
full_bkdir=$bkdir/full_backup
today_bkdir=$full_bkdir/$today
DB_Name=root
DB_PASS=000000
if [ ! -d $bkdir ] ; then
      mkdir -p $bkdir
else
      chattr -i $bkdir
fi
if [ ! -d $full_bkdir ] ; then
mkdir -p $full_bkdir
else
      chattr -i $full_bkdir
fi
if [ ! -d $today_bkdir ] ; then
      mkdir -p $today_bkdir
else
      chattr -i $today_bkdir
fi
database=NOT_NULL
for ((i=2;i<10;i++))
do
      if [ -n "$database" ];then
      database='mysql -u$DB_Name -p$DB_PASS -S /var/lib/mysql/mysql.sock \
-e "show databases;"|sed '1,'$i'd' |sed '2,$d''
      mysqladmin -u$DB_Name -p$DB_PASS flush-tables
      mysqldump -u$DB_Name -p$DB_PASS $database \
-A --events ignore-tables=mysql.events>$today_bkdir/$database.sql
```

```
        cd $full_bkdir
        for d in '"find . -type d -mtime +6 -maxdepth 1">/dev/null 2>&1'
        do
                chattr -i $d
                rm -fr $d
        done
        chattr +i $bkdir
        echo "$database is fine"
        else
        break
        fi
done
```

附录十二　mysql_hourly_bk.sh

数据库定时备份脚本

```
#!/bin/bash
today='date +"%Y_%m_%d"'
bkdir=/opt/mysql/backup/
hourly_bkdir=$bkdir/hourly_backup
today_bkdir=$hourly_bkdir/$today
DB_Name=root
DB_PASS=000000
log_dir=/var/lib/mysql
if [ ! -d $bkdir ] ; then
        mkdir -p $bkdir
else
        chattr -i $bkdir
fi
if [ ! -d $hourly_bkdir ] ; then
        mkdir -p $hourly_bkdir
else
        chattr -i $hourly_bkdir
fi
if [ ! -d $today_bkdir ] ; then
        mkdir -p $today_bkdir
else
        chattr -i $today_bkdir
fi
```

```
mysqladmin -u$DB_Name -p$DB_PASS flush-logs
total='ls $log_dir/mysql-bin.*|wc -l'
total='expr $total - 2'
for f in 'ls $log_dir/mysql-bin.*|head -n $total'
do
        bf='basename $f'
        echo $bf is finish.
        mv $f $today_bkdir
done
```

附录十三　ovs-network.txt

OpenvSwitch 网络配置脚本

```
ovs-vsctl add-br br-mgmt
ovs-vsctl add-br br-eth0
ovs-vsctl add-port br-eth0 eth0
ovs-vsctl add-port br-mgmt "br-mgmt-br-eth0"
ovs-vsctl add-port br-eth0 "br-eth0-br-mgmt"
ovs-vsctl set interface  "br-eth0-br-mgmt" type=patch
ovs-vsctl set interface  "br-mgmt-br-eth0" type=patch
ovs-vsctl set interface "br-eth0-br-mgmt" options:peer="br-mgmt-br-eth0"
ovs-vsctl set interface "br-mgmt-br-eth0" options:peer="br-eth0-br-mgmt"
ovs-vsctl add-br br-eth1
ovs-vsctl add-br br-prv
ovs-vsctl add-br br-int
ovs-vsctl add-br br-ex
ovs-vsctl add-port br-eth1 eth1
ovs-vsctl add-port br-eth1 "br-eth1-br-int"
ovs-vsctl add-port br-eth1 "br-eth1-br-prv"
ovs-vsctl add-port br-eth1 "br-eth1-br-ex"
ovs-vsctl add-port br-prv "br-prv-br-eth1"
ovs-vsctl add-port br-int "br-int-br-eth1"
ovs-vsctl add-port br-ex "br-ex-br-eth1"
ovs-vsctl set interface  "br-prv-br-eth1" type=patch
ovs-vsctl set interface  "br-eth1-br-prv" type=patch
ovs-vsctl set interface  "br-eth1-br-ex" type=patch
ovs-vsctl set interface  "br-ex-br-eth1" type=patch
ovs-vsctl set interface  "br-eth1-br-int" type=patch
ovs-vsctl set interface  "br-int-br-eth1" type=patch
```

```
ovs-vsctl set interface "br-prv-br-eth1" options:peer="br-eth1-br-prv"
ovs-vsctl set interface "br-eth1-br-prv" options:peer="br-prv-br-eth1"
ovs-vsctl set interface "br-eth1-br-ex" options:peer="br-ex-br-eth1"
ovs-vsctl set interface "br-ex-br-eth1" options:peer="br-eth1-br-ex"
ovs-vsctl set interface "br-eth1-br-int" options:peer="br-int-br-eth1"
ovs-vsctl set interface "br-int-br-eth1" options:peer="br-eth1-br-int"
ovs-vsctl set port "br-ex-br-eth1" trunks=0
ovs-vsctl set port "br-eth1-br-ex" trunks=0
```

附录十四　ovs-show.txt

OpenvSwitch 网络信息查看

```
$ ovs-vsctl show
52f4ba6f-4291-482a-a0f7-220ef417eeb8
    Bridge br-ex                        //OpenvSwitch 实例外部访问网桥
        Port "br-ex-br-eth1"            //属于连接 br-eth1 的端口
            trunks: [0]
            Interface "br-ex-br-eth1"
                type: patch
                options: {peer="br-eth1-br-ex"}
        Port br-ex
            Interface br-ex
                type: internal
    Bridge "br-eth0"                    //OpenvSwitch eth0 接口网桥
        Port "br-eth0-br-mgmt"          //属于连接 br-mgmt 的端口
            Interface "br-eth0-br-mgmt"
                type: patch
                options: {peer="br-mgmt-br-eth0"}
        Port "eth0"
            Interface "eth0"
        Port "br-eth0"
            Interface "br-eth0"
                type: internal
    Bridge br-mgmt                      //OpenvSwitch 管理网络
        Port br-mgmt
            Interface br-mgmt
                type: internal
        Port "br-mgmt-br-eth0"          //属于连接 br-eth0 的端口
            Interface "br-mgmt-br-eth0"
```

```
            type: patch
            options: {peer="br-eth0-br-mgmt"}
  Bridge br-prv                      //OpenvSwitch 实例私有用户网络
      Port br-prv
          Interface br-prv
              type: internal
      Port "br-prv-br-eth1"          //属于连接 br-eth1 的端口
          Interface "br-prv-br-eth1"
              type: patch
              options: {peer="br-eth1-br-prv"}
      Port phy-br-prv
          Interface phy-br-prv
  Bridge "br-eth1"                   //OpenvSwitch eth1 网桥
      Port "br-eth1-br-ex"           //属于连接 br-ex 的端口
          trunks: [0]
          Interface "br-eth1-br-ex"
              type: patch
              options: {peer="br-ex-br-eth1"}
      Port "br-eth1-br-prv"          //属于连接 br-prv 的端口
          Interface "br-eth1-br-prv"
              type: patch
              options: {peer="br-prv-br-eth1"}
      Port "br-eth1"
          Interface "br-eth1"
              type: internal
      Port "eth1"
          Interface "eth1"
      Port "br-eth1-br-int"          //属于连接 br-eth1 的端口
          Interface "br-eth1-br-int"
              type: patch
              options: {peer="br-int-br-eth1"}
  Bridge br-int                      //OpenvSwitch 实例接入网桥
      fail_mode: secure
      Port "br-int-br-eth1"          //属于连接 br-eth1 的端口
          Interface "br-int-br-eth1"
              type: patch
              options: {peer="br-eth1-br-int"}
      Port int-br-prv
```

```
            Interface int-br-prv
        Port br-int
            Interface br-int
                type: internal
    ovs_version: "2.1.3"
```

附录十五　environment.txt

基础环境配置脚本

```
source $(pwd)/Xiandian_Pre.sh                           // 生效环境配置文件
service NetworkManager stop 2>/dev/null                 //关闭网络管理服务
chkconfig NetworkManager off 2>/dev/null
setenforce 0  2>&1                                      //修改 selinux
yum upgrade -y                                          //升级系统
iptables -F
iptables -X
iptables -Z
service iptables save
sed -i "/REJECT/i\-A INPUT -m state --state NEW -m tcp -p tcp --dport 27017 \
-j ACCEPT" /etc/sysconfig/iptables
sed -i "/INPUT.*REJECT/i\-A INPUT -m state --state NEW -m tcp -p tcp --dport
5900:5909 \
-j ACCEPT" /etc/sysconfig/iptables
service iptables reload                                 //清除防火墙规则保存配置
sed -i 's/SELINUX=enforcing/SELINUX=permissive/g' /etc/selinux/config
cat <<EOF >> /etc/hosts
$Controller_Mgmt_IPAddress  $Contoller_Hostname
$Network_Mgmt_IPAddress     $Network_Hostname
$Compute_Mgmt_IPAddress     $Compute_Hostname
$Stroage_Mgmt_IPAddress     $Stroage_Hostname
EOF
// 修改 hosts 文件，加入主机解析
ssh-keygen
yum install -y openstack-utils openssh-clients openstack-selinux openssh-
clients perl wget \
openssh-clients expect ntp openvswitch mysql mysql-server MySQL-python expect \
openstack-keystone python-keystoneclient  \
//安装依赖软件包
```

```
for  hosts in $Contoller_Hostname $Network_Hostname $Compute_Hostname $Stroage_
Hostname; \
do ssh-copy-id -i /root/.ssh/id_rsa.pub $hosts  && ssh $hosts "yum -y install
openssh-clients";done
cat > /etc/sysconfig/network-scripts/ifcfg-br-mgmt  << EOF
DEVICE=br-mgmt
IPADDR=$Controller_Mgmt_IPAddress
NETMASK=255.255.255.0
BOOTPROTO=static
GATEWAY=$Gateway_Mgmt
ONBOOT=yes
USERCTL=no
EOF                                       //修改网卡默认配置地址
```

附录十六　mysql.txt

数据库安装配置

```
sed  -i   "/^symbolic-links/a\default-storage-engine  =  innodb\ninnodb_file_
per_table\ncollation-server = \
utf8_general_ci\ninit-connect = 'SET NAMES utf8'\ncharacter-set-server = utf8"
/etc/my.cnf
service mysqld start
chkconfig mysqld on
expect -c "
spawn /usr/bin/mysql_secure_installation
expect \"Enter current password for root (enter for none):\"
send \"\r\"
expect \"Set root password?\"
send \"y\r\"
expect \"New password:\"
send \"$Mysql_Admin_Passwd\r\"
expect \"Re-enter new password:\"
send \"$Mysql_Admin_Passwd\r\"
expect \"Remove anonymous users?\"
send \"y\r\"
expect \"Disallow root login remotely?\"
send \"n\r\"
expect \"Remove test database and access to it?\"
send \"y\r\"
```

```
expect \"Reload privilege tables now?\"
send \"y\r\"
expect eof
"
//创建数据库用户和用户密码
mysql -uroot -p$Mysql_Admin_Passwd -e "create database IF NOT EXISTS keystone ;"
mysql -uroot -p$Mysql_Admin_Passwd -e "GRANT ALL PRIVILEGES ON keystone.* \
TO 'keystone'@'localhost' IDENTIFIED BY '$Demo_DB_Passwd' ;"
......
//配置每个服务的访问连接地址和赋予访问用户的权限
sed -i "s/bind_ip.*/bind_ip = 127.0.0.1,$Controller_Mgmt_IPAddress/g" /etc
/mongodb.conf
//设置 mongodb 的默认监听地址
echo "smallfiles = true">>/etc/mongodb.conf
service mongod restart
chkconfig mongod on
echo -e "\033[31m   Please waiting......        \033[0m "
sleep 90
mongo $Controller_Mgmt_IPAddress/ceilometer --eval "db.addUser({user: 'ceilometer',
pwd: '$Demo_DB_Passwd', roles: [ 'readWrite', 'dbAdmin' ]})"
while [ $? -ne 0 ]
do
sleep 60
echo "Please waiting......"
mongo $Controller_Mgmt_IPAddress/ceilometer \
--eval  "db.addUser({user:  'ceilometer',  pwd:  '$Demo_DB_Passwd',  roles:
[ 'readWrite', 'dbAdmin' ]})"
done
//创建 mongo 默认的数据库的用户和密码
# Configure databases connectios
openstack-config --set /etc/keystone/keystone.conf database \
connection mysql://keystone:$Demo_DB_Passwd@$Controller_Mgmt_IPAddress/keystone
......
//配置服务的数据库连接地址
# Update Services Tables
su -s /bin/sh -c "keystone-manage db_sync" keystone
su -s /bin/sh -c "glance-manage db_sync" glance
su -s /bin/sh -c "nova-manage db sync" nova
```

```
su -s /bin/sh -c "cinder-manage db sync" cinder
su -s /bin/sh -c "heat-manage db_sync" heat
// 同步创建数据库表
//配置基本环境之后开始每个服务自己的配置过程
//包括 keystone 认证信息的配置、每个服务的认证息的用户、服务、访问端点的配置，如下所示。
keystone user-create --name=glance --pass=$Demo_User_Passwd
keystone user-role-add --user=glance --tenant=service --role=admin
keystone service-create --name=glance --type=image --description= "OpenStack
Image Service"
keystone endpoint-create --service-id=$(keystone service-list | awk '/ image /
{print $2}') \
--publicurl=http://$Controller_External_IPAddress:9292 \
--internalurl=http://$Controller_Mgmt_IPAddress:9292 \
--adminurl=http://$Controller_Mgmt_IPAddress:9292
......
//创建服务的用户名和服务，同时创建服务的访问 api 端点
......
perl -p -i -e "s/^(if _fastmath is not None .*:)/#\1/" /usr/lib64/python2.6/
site-packages/Crypto/Util/number.py
perl -p -i -e "s/^(\s*_warn.*Not using mpz_powm_sec.*)/#\1/" \
/usr/lib64/python2.6/site-packages/Crypto/Util/number.py
openstack-config --set /etc/glance/glance-api.conf keystone_authtoken \
auth_uri http://$Controller_Mgmt_IPAddress:5000
openstack-config --set /etc/glance/glance-api.conf keystone_authtoken auth_
host $Controller_Mgmt_IPAddress
openstack-config --set /etc/glance/glance-api.conf keystone_authtoken auth_
port 35357
openstack-config --set /etc/glance/glance-api.conf keystone_authtoken auth_
protocol http
openstack-config --set /etc/glance/glance-api.conf keystone_authtoken admin_
tenant_name service
openstack-config --set /etc/glance/glance-api.conf keystone_authtoken admin_
user glance
openstack-config --set /etc/glance/glance-api.conf keystone_authtoken admin_
password $Demo_User_Passwd
openstack-config --set /etc/glance/glance-api.conf paste_deploy flavor keystone
openstack-config --set /etc/glance/glance-api.conf DEFAULT notification_driver
messaging
```

```
openstack-config --set /etc/glance/glance-registry.conf keystone_authtoken \
auth_uri http://$Controller_Mgmt_IPAddress:5000
openstack-config --set /etc/glance/glance-registry.conf keystone_authtoken \
auth_host $Controller_Mgmt_IPAddress
openstack-config --set /etc/glance/glance-registry.conf keystone_authtoken auth_
port 35357
openstack-config --set /etc/glance/glance-registry.conf keystone_authtoken auth_
protocol http
openstack-config  --set  /etc/glance/glance-registry.conf  keystone_authtoken
admin_tenant_name service
openstack-config  --set  /etc/glance/glance-registry.conf  keystone_authtoken
admin_user glance
openstack-config --set /etc/glance/glance-registry.conf keystone_authtoken \
admin_password $Demo_User_Passwd
openstack-config --set /etc/glance/glance-registry.conf paste_deploy flavor
keystone
openstack-config --set /etc/glance/glance-api.conf DEFAULT rpc_backend qpid
......
//配置服务的 Keystone 认证信息，配置每个服务组件信息之后，批量启动服务并设置为开机自启
chkconfig openstack-glance-api on
......
service openstack-glance-api restart
......
```

附录十七 compute.txt

计算节点安装配置

```
openstack-config --set /etc/nova/nova.conf DEFAULT auth_strategy keystone
//配置认证
......
openstack-config --set /etc/nova/nova.conf DEFAULT rpc_backend qpid
openstack-config  --set  /etc/nova/nova.conf  DEFAULT  qpid_hostname  $Controller_
Mgmt_IPAddress
openstack-config  --set  /etc/nova/nova.conf  DEFAULT  my_ip  $Compute_Mgmt_
IPAddress
openstack-config --set /etc/nova/nova.conf DEFAULT vnc_enabled True
openstack-config --set /etc/nova/nova.conf DEFAULT vncserver_listen 0.0.0.0
openstack-config --set /etc/nova/nova.conf DEFAULT \
vncserver_proxyCLIent_address $Compute_Mgmt_IPAddress
```

```
openstack-config --set /etc/nova/nova.conf DEFAULT \
novncproxy_base_url http://$Controller_Mgmt_IPAddress:6080/vnc_auto.html
//配置 VNC 服务
openstack-config --set /etc/nova/nova.conf DEFAULT glance_host $Controller_
Mgmt_IPAddress
sed -i "/INPUT.*REJECT/i\-A INPUT -m state --state NEW -m tcp -p tcp --dport
5900:5909 \
-j ACCEPT" /etc/sysconfig/iptables
openstack-config --set /etc/nova/nova.conf DEFAULT instance_usage_audit True
openstack-config --set /etc/nova/nova.conf DEFAULT instance_usage_ audit_period
hour
openstack-config --set /etc/nova/nova.conf DEFAULT notify_on_state_change
vm_and_task_state
openstack-config --set /etc/nova/nova.conf DEFAULT \
notification_driver nova.openstack.common.notifier.rpc_notifier
openstack-config --set /etc/nova/nova.conf DEFAULT nnotification_driver
ceilometer.compute.nova_notifier
sed -i 's/nnotification_driver/notification_driver/g' /etc/nova/nova.conf
openstack-config --set /etc/nova/nova.conf DEFAULT rpc_backend qpid
//配置监控和消息服务
openstack-config --set /etc/nova/nova.conf DEFAULT network_api_class nova.
network.neutronv2.api.API
openstack-config --set /etc/nova/nova.conf DEFAULT neutron_url http://
$Controller_Mgmt_IPAddress:9696
openstack-config --set /etc/nova/nova.conf DEFAULT neutron_auth_strategy
keystone
openstack-config --set /etc/nova/nova.conf DEFAULT neutron_admin_tenant_ name
service
openstack-config --set /etc/nova/nova.conf DEFAULT neutron_admin_username
neutron
openstack-config --set /etc/nova/nova.conf DEFAULT neutron_admin_password
$Demo_User_Passwd
openstack-config --set /etc/nova/nova.conf DEFAULT \
neutron_admin_auth_url http://$Controller_Mgmt_IPAddress:35357/v2.0
openstack-config --set /etc/nova/nova.conf DEFAULT \
linuxnet_interface_driver nova.network.linux_net.LinuxOVSInterfaceDriver
openstack-config --set /etc/nova/nova.conf DEFAULT firewall_driver nova.
virt.firewall.NoopFirewallDriver
```

```
openstack-config --set /etc/nova/nova.conf DEFAULT security_group_api neutron
//配置实例使用 Neutron 网络
openstack-config --set /etc/nova/nova.conf DEFAULT vif_plugging_is_fatal False
openstack-config --set /etc/nova/nova.conf DEFAULT vif_plugging_timeout 10
openstack-config --set /etc/nova/nova.conf DEFAULT service_neutron_metadata_
proxy true
openstack-config --set /etc/nova/nova.conf DEFAULT neutron_metadata_ proxy_
shared_secret 000000
openstack-config --set /etc/nova/nova.conf DEFAULT instance_usage_audit True
openstack-config --set /etc/nova/nova.conf DEFAULT instance_usage_audit_ period
hour
openstack-config --set /etc/nova/nova.conf DEFAULT notify_on_state_change
vm_and_task_state
sed -i 's/notification_driver.*//g' /etc/nova/nova.conf
openstack-config --set /etc/nova/nova.conf DEFAULT \
notification_driver nova.openstack.common.notifier.rpc_notifier
sed -i 's/notification_driver.*/notification_driver = \
nova.openstack.common.notifier.rpc_notifier\nnotification_driver = ceilometer.
compute. nova_notifier/g' \
/etc/nova/nova.conf
```